Ulrich Preiss

Quantum Chemical Studies of Weakly Coordinated Ionic Systems

Ulrich Preiss

Quantum Chemical Studies of Weakly Coordinated Ionic Systems

Predictions of Chemical and Physical Properties

Südwestdeutscher Verlag für Hochschulschriften

Impressum/Imprint (nur für Deutschland/ only for Germany)
Bibliografische Information der Deutschen Nationalbibliothek: Die Deutsche Nationalbibliothek verzeichnet diese Publikation in der Deutschen Nationalbibliografie; detaillierte bibliografische Daten sind im Internet über http://dnb.d-nb.de abrufbar.

Alle in diesem Buch genannten Marken und Produktnamen unterliegen warenzeichen-, marken- oder patentrechtlichem Schutz bzw. sind Warenzeichen oder eingetragene Warenzeichen der jeweiligen Inhaber. Die Wiedergabe von Marken, Produktnamen, Gebrauchsnamen, Handelsnamen, Warenbezeichnungen u.s.w. in diesem Werk berechtigt auch ohne besondere Kennzeichnung nicht zu der Annahme, dass solche Namen im Sinne der Warenzeichen- und Markenschutzgesetzgebung als frei zu betrachten wären und daher von jedermann benutzt werden dürften.

Verlag: Südwestdeutscher Verlag für Hochschulschriften GmbH & Co. KG
Dudweiler Landstr. 99, 66123 Saarbrücken, Deutschland
Telefon +49 681 37 20 271-1, Telefax +49 681 37 20 271-0
Email: info@svh-verlag.de
Zugl.: Freiburg, Universität, Dissertation, 2010

Herstellung in Deutschland:
Schaltungsdienst Lange o.H.G., Berlin
Books on Demand GmbH, Norderstedt
Reha GmbH, Saarbrücken
Amazon Distribution GmbH, Leipzig
ISBN: 978-3-8381-2064-5

Imprint (only for USA, GB)
Bibliographic information published by the Deutsche Nationalbibliothek: The Deutsche Nationalbibliothek lists this publication in the Deutsche Nationalbibliografie; detailed bibliographic data are available in the Internet at http://dnb.d-nb.de.

Any brand names and product names mentioned in this book are subject to trademark, brand or patent protection and are trademarks or registered trademarks of their respective holders. The use of brand names, product names, common names, trade names, product descriptions etc. even without a particular marking in this works is in no way to be construed to mean that such names may be regarded as unrestricted in respect of trademark and brand protection legislation and could thus be used by anyone.

Publisher: Südwestdeutscher Verlag für Hochschulschriften GmbH & Co. KG
Dudweiler Landstr. 99, 66123 Saarbrücken, Germany
Phone +49 681 37 20 271-1, Fax +49 681 37 20 271-0
Email: info@svh-verlag.de

Printed in the U.S.A.
Printed in the U.K. by (see last page)
ISBN: 978-3-8381-2064-5

Copyright © 2010 by the author and Südwestdeutscher Verlag für Hochschulschriften GmbH & Co. KG and licensors
All rights reserved. Saarbrücken 2010

Table Of Contents

1. Prologue 5
2. Methodology 6
   2.1. General Computational and Methodological Details 6
      2.1.1. Error handling 6
      2.1.2. Quantum chemical calculations 6
      2.1.3. Nomenclature of Ionic Liquids 9
      2.1.4. The Conformational Space 11
   2.2. Molecular Structures 18
      2.2.1. Calculated Weakly Coordinating Anions and Related Species 18
      2.2.2. Calculated Ion Pairs 32
      2.2.3. Experimental Structures 39
3. Weakly Coordinating Anions 51
   3.1. General Introduction 51
   3.2. Computational Study of $[M(CF_3)_4]^-$-type WCAs 52
      3.2.1. Background 52
      3.2.2. Results 52
      3.2.3. Conclusion 60
      3.2.4. Appendix: Additional Computational Details 60
   3.3. Computational Screening of Further Possible Weakly-Coordinating Anions 62
      3.3.1. Background 62
      3.3.2. Results 65
      3.3.3. Conclusion 68
4. Ionic Liquids 70
   4.1. General Introduction 70
      4.1.1. The Near-Ordering of Ionic Liquids: Servants of two Masters 70
   4.2. The Molecular Volume 72
      4.2.1. Background 74
      4.2.2. Results 75
      4.2.3. Conclusion 77
      4.2.4. Appendix: Additional Computational Details 78
   4.3. The Temperature-Dependent Density 79
   4.4. The Integrated Heat Capacity 83

- 4.5. The Temperature-Dependent Liquid Entropy 86
  - 4.5.1. Background 86
  - 4.5.2. Results 86
  - 4.5.3. Conclusion 90
- 4.6. Basic Phase Change Thermodynamics 91
  - 4.6.1. The Dissociation Enthalpy 92
  - 4.6.2. The Vaporization Enthalpy 95
  - 4.6.3. The Fusion Enthalpy 99
  - 4.6.4. The Solvation Enthalpy 100
  - 4.6.5. The Lattice Enthalpy 104
  - 4.6.6. Conclusion 109
- 4.7. The Melting Point 111
  - 4.7.1. Background 113
  - 4.7.2. Results 116
  - 4.7.3. Conclusion 122
- 4.8. The Glass Transition Temperature 123
  - 4.8.1. Background 123
  - 4.8.2. Results 127
- 4.9. The Critical Micelle Concentration 133
  - 4.9.1. Background 133
  - 4.9.2. Results 134
  - 4.9.3. Conclusion 140
- 4.10. Further Considerations about IL Nanostructures 142
- 5. Experimental part 144
  - 5.1. Experimental Techniques 144
    - 5.1.1. General Procedures and Starting Materials 144
    - 5.1.2. Analytical Methods 144
  - 5.2. Synthesis and Characterization of the Prepared Compounds 145
    - 5.2.1. Synthesis of Li[Nb(hfip)$_6$] 145
    - 5.2.2. Synthesis of Li[Ta(hfip)$_6$] 148
    - 5.2.3. Synthesis of Ag[Nb(hfip)$_6$] 151
    - 5.2.4. Synthesis of Na-pfad 153
    - 5.2.5. Synthesis of Na[Al(pfad)$_4$] 155

- 5.2.6. Crystallization of ILs 158
- 5.2.7. Crystal Structures 159
6. Abbreviations 170
7. General Conclusion and Outlook 171
   - 7.1. Weakly Coordinating Anions 171
   - 7.2. Ionic Liquids 172
8. References 175

1. Prologue

The advent of faster computers and the decline of hardware costs since the 1950s have greatly facilitated the implementation of quantum chemical methods into everyday chemistry and helped to spread the use of the computer in laboratories around the world not only as a tool to answer theoretical questions, but also to enable the planning of reactions in order to save time, effort and money beforehand, not to speak of the greatly reduced overall risk that's intrinsic to every synthesis. Faster computers also fueled the perpetual development and high availability of better quantum chemical theories.

While for many purposes, strongly bonded systems could be treated satisfyingly since the invention of the Hartree-Fock method,[1,2] weak bonds are much harder to describe as the magnitude of their strength is overshadowed by the error made in the calculative approximations. Amongst weakly bonded systems are weakly coordinating anions (WCAs) and ionic liquids (ILs), compounds of increasing preparative and theoretical interest. In this study, it will be demonstrated amongst other things that in the cases in which a weak bond cannot be completely neglected, a purely statistical treatment often suffices, for example, to describe bulk physical properties, but also to find new proof and new perspectives for theories that were postulated a long time ago.

2. Methodology

2.1. General Computational and Methodological Details

Ordinary least-squares fits (OLS) and some other basic calculations were accomplished with the program QTOctave.[3]

2.1.1. Error handling

The mean error (err$_\varnothing$) is defined as:

$$(1) \quad \text{err}_\varnothing := \frac{\sum_i \left| \frac{x_i}{y_i} - 1 \right|}{N} \cdot 100\%,$$

where y denotes the independent observable and x the calculated value. Likewise, the root mean square error (rmse) is defined as:

$$(2) \quad \text{rmse} := \sqrt{\frac{\sum_i (y_i - x_i)^2}{N}}$$

The error in terms of standard deviation is defined as:

$$(3) \quad \sigma := \sqrt{\frac{\sum_i (x_i - \bar{x})^2}{N-1}}$$

The error of the estimate of two series x and y is defined as:

$$(4) \quad \sigma_{est} := \sqrt{\frac{\sum_i (y_i - x_i)^2}{N-2}} = \text{rmse} \cdot \sqrt{\frac{N}{N-2}}$$

The correlation coefficient r is defined as:

$$(5) \quad r = \frac{\sum_i (x_i - \bar{x})(y_i - \bar{y})}{\sqrt{\left[\sum_i (x_i - \bar{x})^2\right]\left[\sum_i (y_i - \bar{y})^2\right]}}$$

However, the squared expression (r^2) was used throughout this study.

2.1.2. Quantum chemical calculations

From reasonable starting structures, BP86/SV(P)[4,5,6,7] optimizations were carried out with the TURBOMOLE program package (version 5.10)[8] using the resolution of identity (RI) approximation.[9] From BP86/SV(P) geometries, vibrational frequencies were calculated with

AOFORCE for each molecule or molecular ion to make sure they represent a true minimum.[10,11] All gas phase entropies (S_g) and zero-point energies (E_{ZP}) were also taken from these computations and freeh, a module for statistical thermodynamics. Zero-point energies were used unscaled, because the scaling factors for BP86/SV(P) are almost unity.

If applicable, these geometries were then used for further optimization with the TZVP basis set,[12] after which a full optimization with COSMO[13] (using optimized radii; dielectric constant set to infinity) was appended. The geometries of the ion pairs could in some cases change radically during this optimization. For the largest molecules ($\geq$ ['BHF']$^-$), only a single-point calculation with BP86/TZVP and COSMO was performed. All gas phase structures were calculated in the highest possible point group; COSMO calculations were always done without symmetry (C_1). The output files were then read in by COSMOtherm using the BP_TZVP_C21_0106 parametrization.[14]

The geometries of IL species are accessible via ssh from the intranet at the University of Freiburg at the IP address 132.230.120.184 in the directory /share/ilgroup and its subdirectories. A total of 221 cations (834 MB), 127 anions (346 MB), and 97 ion pairs (839 MB) were calculated. Of these, the conformer space of 74 cations, 24 anions, and 3 ion pairs was sampled with 2...99 conformers each, amounting to a total size of 15 GB. For WCA calculations, see section 2.2.1.

Background on COSMO and COSMO-RS

COSMO, the "COnductor-like Screening MOdel" approach, belongs to the continuum solvation models, i.e. the electrostatic behavior of a solvent is described by a dielectric continuum. Instead of using the exact dielectric boundary condition, COSMO applies the simpler boundary condition of the vanishing potential on the surface of a conducting medium. The polarization of the continuum, induced by the charge density of the solute, is represented by the screening charge density σ appearing on the continuum solvent boundary surface and can be calculated by solving the boundary condition problem. The used cavity is of molecular shape consisting of sufficiently small segments leading to a discretization of the problem. The procedure can be implemented in self-consistent field (SCF) calculations (Hartree-Fock or Kohn-Sham). In order to cross over to a dielectric continuum, the screening charge density σ can be scaled down to a finite dielectric constant ε_r.

COSMO-RS ("COSMO for Real Solvents") is a predictive method for the calculation of thermodynamic properties of fluids that uses a statistical thermodynamics approach based on the results of COSMO SCF calculations for molecules embedded in an electrical conductor, i.e. using $\varepsilon_r = \infty$. The liquid can be imagined as a dense packing of molecules in the described reference state. For the statistical thermodynamic procedure, this system is broken down to an ensemble of pair-wise interacting surface segments. The interaction can be expressed in terms of surface descriptors, whereof σ is the most important one. The interaction energy modes, i.e. electrostatics (H_{MF}) and hydrogen bonding (H_{HB}), are described as functions of the screening charge densities of two interacting surface segments σ and σ'. The misfit enthalpy describes the interaction of surface areas with different polarities. The less specific van-der-Waals (H_{vdW}) interactions are taken into account in a slightly more approximate way by element-specific interaction terms. A more detailed description, which is beyond the scope of this thesis, can be found in ref. [15].

Common descriptors

Apart from these enthalpies, several basic quantities are used throughout this study. The solvent-accessible surface, denoted as $\hat{S}$, can be output out by several programs; in this case, we employed COSMOtherm.[14] The molecular volume (V_m) is a measure of the size of a molecule; it is calibrated in order to reconstruct a single unit cell of a certain compound in the lattice and, like the surface, contains information about the number of interactions a molecule can form. Details are given further down (section 4.2). Another measure for size is the molecular radius, r_m. It is the sum of anionic (r_{ion}^-) and cationic radius (r_{ion}^+) which are both calculated from ionic volumes assuming ideally spherical behavior:

$$(6) \quad r_{ion}^\pm = \sqrt[3]{\frac{3V_{ion}^\pm}{4\pi}}$$

The ratio of V_m to r_m^3 is almost 1 (π/3, to be exact), if cationic and anionic radii are identical ($r_m^+ = r_m^-$), and gets larger the more they differ, until it quadruples when one radius vanishes. Therefore, r_m^3 could be better suited for describing the compound-specific interactions for molecules where cation and anion strongly differ in size. Figure 1 is a schematical depiction of this relationship.

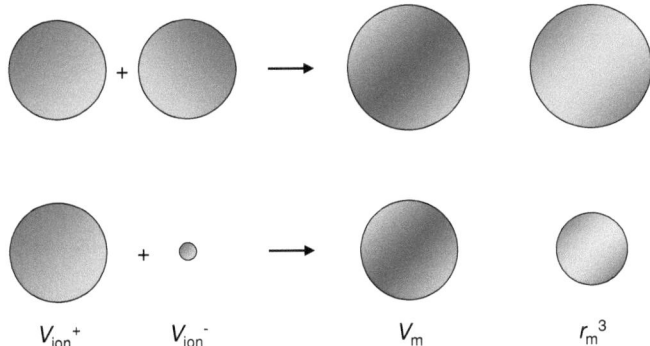

Figure 1: The relation of V_m and r_m^3 for spherical ions.

2.1.3. Nomenclature of Ionic Liquids

For ammonium, phosphonium and sulfonium cations $[E_{n,n,n(,n)}]^+$ (E = N, P, S), the comma-separated indices show the number of C-atoms the linear alkyl ligands possess; the index φ stands for phenyl, i3 for isopropyl. Complex anions are described in Table 1, complex cations in Table 2.

As for abbreviated compounds: ECOENG 41M = $[C_4MIm][CH_3OC_2H_4OC_2H_4SO_4]$, ECOENG 500 = $[Cat][C_1SO_4]$, where the cation is depicted in Figure 2.

Table 1: Complex anions.

anion	formula	anion	formula
$[C_nSO_3]^-$	$[C_nH_{2n+1}SO_3]^-$	[NTfaTf]⁻	$[N(C(O)CF_3)(SO_2CF_3)]^-$
$[C_nSO_4]^-$	$[C_nH_{2n+1}SO_4]^-$	[OAc]⁻	$[CH_3CO_2]^-$
$[CTf_3]^-$	$[C(SO_2CF_3)_3]^-$	[OTf]⁻	$[CF_3SO_3]^-$
[Dca]⁻	$[N(CN)_2]^-$	[OTfa]⁻	$[CF_3CO_2]^-$
[Doc]⁻	(structure)	[OTos]⁻	(structure)
[Fap]⁻	$[P(C_2F_5)_3F_3]^-$	[Tcm]⁻	$[C(CN)_3]^-$
$[NFs_2]^-$	$[N(SO_2F)_2]^-$	['HF']⁻	$[Al(OC(CF_3)_2H)_4]^-$
$[NNf_2]^-$	$[N(SO_2C_4F_9)_2]^-$	['BHF']⁻	$[B(OC(CF_3)_2H)_4]^-$
$[NPf_2]^-$	$[N(SO_2C_2F_5)_2]^-$	['HT']⁻	$[Al(OC(CF_3)_2CH_3)_4]^-$
$[NTf_2]^-$	$[N(SO_2CF_3)_2]^-$	['PF']⁻	$[Al(OC(CF_3)_3)_4]^-$

Table 2: Substituted cations.

cation	structure	R^1	R^2	R^3	R^4	R^5
pyridinium:	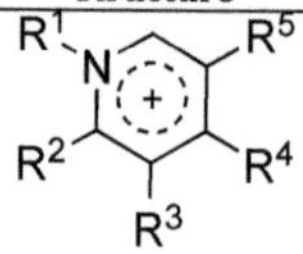 (pyridinium ring with R^1 on N, R^2, R^3, R^4, R^5)					
$[HPy]^+$		H	H	H	H	H
$[C_nPy]^+$		C_nH_{2n+1}	H	H	H	H
$[C_1(CN)Py]^+$		$NC-CH_2$	H	H	H	H
$[C_nC_1Py]^+$		C_nH_{2n+1}	H	CH_3	H	H
$[(C_2)_2Nic]^+$		C_2H_5	H	$C_2H_5OC(O)$	H	H
$[(C_4)_2Nic]^+$		C_4H_9	H	$C_4H_9OC(O)$	H	H
$[C_6DmaPy]^+$		C_6H_{13}	H	H	$N(CH_3)_2$	H
$[C_6C_1DmaPy]^+$		C_6H_{13}	H	CH_3	$N(CH_3)_2$	H
$[C_6(MPip)Py]^+$		C_6H_{13}	H	H	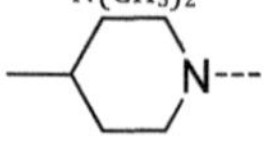 (N-methylpiperidinyl)	H
$[C_6(C_1)_2Py]^+$		C_6H_{13}	H	CH_3	H	CH_3
$[C_6C_2(C_1)_2Py]^+$		C_6H_{13}	C_2H_5	CH_3	H	CH_3
$[C_6C_3(C_2)_2Py]^+$		C_6H_{13}	C_3H_7	C_2H_5	H	C_2H_5
imidazolium:	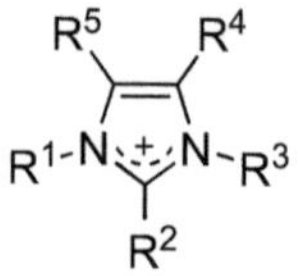 (imidazolium ring with R^1, R^2, R^3, R^4, R^5)	R^1	R^2	R^3	R^4	R^5
$[C_nMIm]^+$		CH_3	H	C_nH_{2n+1}	H	H
$[C_nEIm]^+$		C_2H_5	H	C_nH_{2n+1}	H	H
$[C_nPIm]^+$		C_3H_7	H	C_nH_{2n+1}	H	H
$[C_{i3}MIm]^+$		$i\text{-}C_3H_7$	H	CH_3	H	H
$[C_n(CN)MIm]^+$		CH_3	H	$NC\text{-}(CH_2)_n$	H	H
$[AllyMIm]^+$		CH_3	H	$CH_2CH\text{-}CH_2$	H	H
$[nonafluoro\text{-}C_6MIm]^+$		CH_3	H	$F(CF_2)_4(CH_2)_2$	H	H
$[\varphi CCCMIm]^+$		CH_3	H	$C_6H_5(CH_2)_3$	H	H
$[C_nC_1MIm]^+$		CH_3	CH_3	C_nH_{2n+1}	H	H
$[(C_2)_2EIm]^+$		C_2H_5	C_2H_5	C_2H_5	H	H
$[(C_1)_4MIm]^+$		CH_3	CH_3	CH_3	CH_3	CH_3
pyrrolidinium:	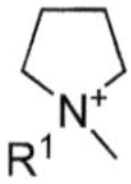 (pyrrolidinium ring with R^1 and methyl)	R^1				
$[C_nMPyr]^+$		C_nH_{2n+1}				
$[C_1(CN)MPyr]^+$		$NC\text{-}CH_2$				
morpholinium:						
$[C_4MMorph]^+$	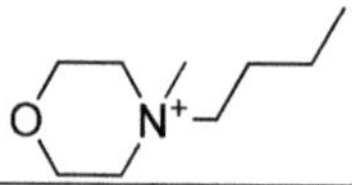 (N-butyl-N-methylmorpholinium structure)					

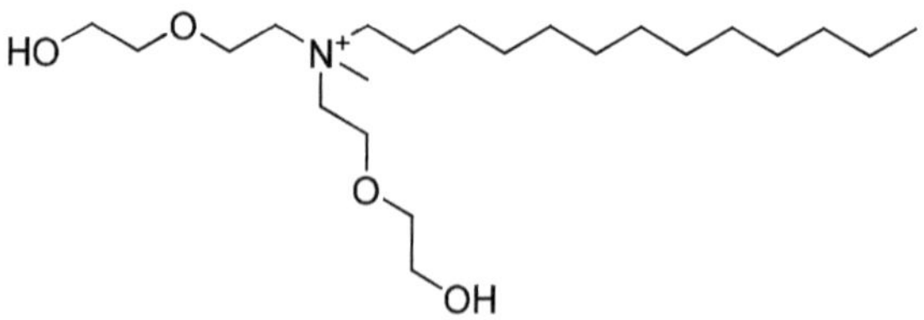

Figure 2: The ECOENG 500 cation $[Cat]^+$.

2.1.4. The Conformational Space

For a representative set of five anions and cations commonly used in ILs, we investigated whether the use of the entire conformational space would make a difference to only using the most stable isomer. In particular, we used $[S_{2,2,2}]^+$, $[C_4MIm]^+$, $[C_4MPyr]^+$, $[C_4Py]^+$, and $[N_{1,1,1,4}]^+$ in combination with $[NTf_2]^-$, $[PF_6]^-$, $[Dca]^-$, $[Fap]^-$, and $[OTf]^-$. For this, conformers were generated with the program rotate,[16,17] then gas phase optimizations with PM6, BP86/SV(P) and BP86/TZVP were performed. The BP86/SV(P) structures were also checked for imaginary frequencies. After each stage, geometrically similar compounds (rmse ≤ 0.05 nm) were identified with the program theseus and reduced to the energetically more favored structure.[18] In the end, optimizations with BP86/TZVP+COSMO were performed. The resulting number of conformers and their energy ranges are given in Table 3; structures are depicted in Figure 3 to Figure 10. Results of COSMO-RS calculations are given for the most important interaction enthalpies which further thermodynamic quantities depend on (see Table 4 to Table 6).

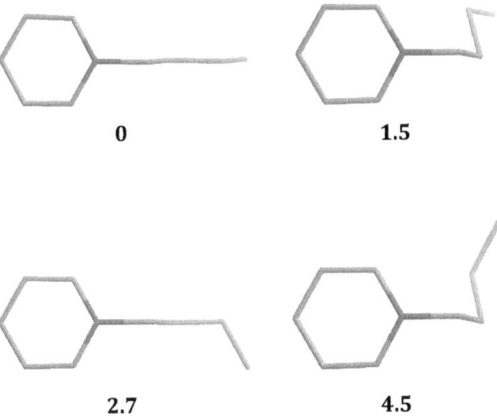

Figure 3: The different conformers of $[C_4Py]^+$, calculated at BP86/TZVP level and sorted by ascending energy. Hydrogen atoms omitted for clarity. Subtitled: the gas phase energies in kJ mol^{-1} relative to the most stable conformer.

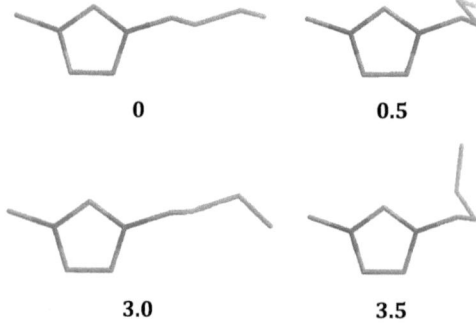

Figure 4: The different conformers of [C$_4$MIm]$^+$, calculated at BP86/TZVP level and sorted by ascending energy. Hydrogen atoms omitted for clarity. Subtitled: the gas phase energies in kJ mol^{-1} relative to the most stable conformer.

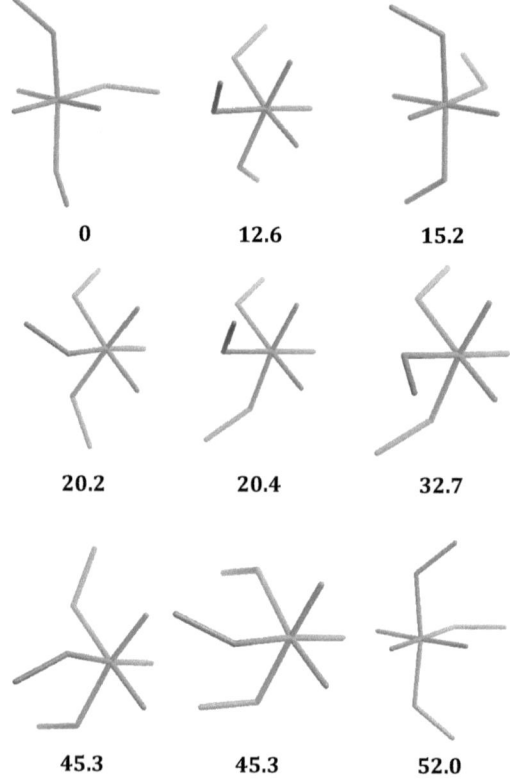

Figure 5: The different conformers of [Fap]$^-$, calculated at BP86/TZVP level and sorted by ascending energy. The fluorine atoms of the alkyl chains were omitted for clarity. Subtitled: the gas phase energies in kJ mol^{-1} relative to the most stable conformer.

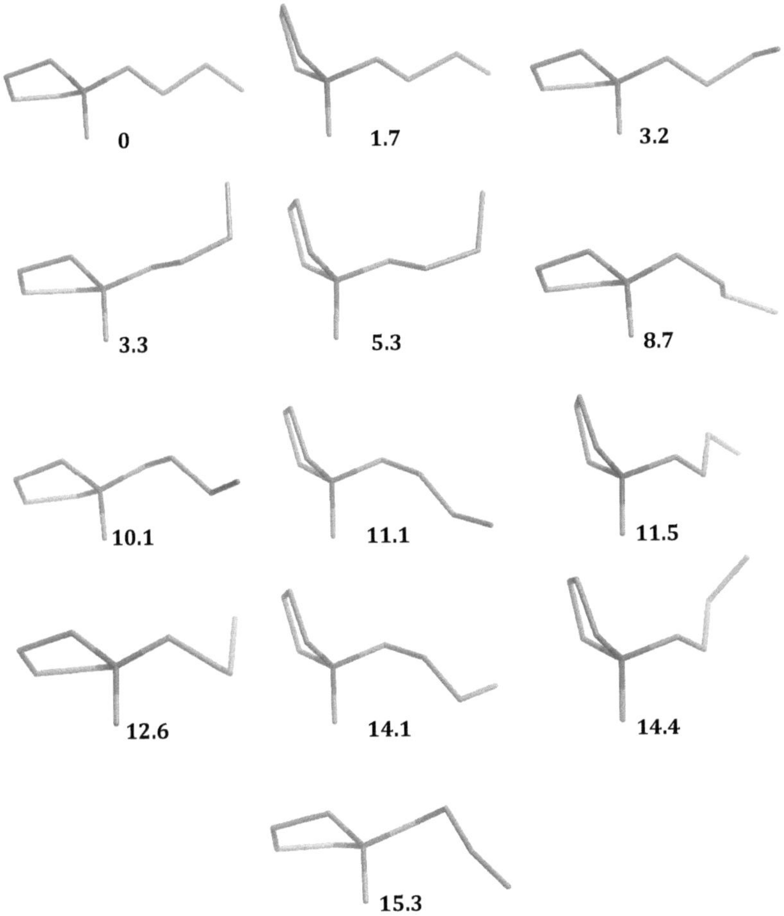

0	1.7	3.2
3.3	5.3	8.7
10.1	11.1	11.5
12.6	14.1	14.4
	15.3	

Figure 6: The different conformers of [C$_4$MPyr]$^+$, calculated at BP86/TZVP level and sorted by ascending energy. Hydrogen atoms omitted for clarity. Subtitled: the gas phase energies in kJ mol^{-1} relative to the most stable conformer.

0   0.1

Figure 7: The different conformers of [NTf$_2$]$^+$, calculated at BP86/TZVP level and sorted by ascending energy. Subtitled: the gas phase energies in kJ mol^{-1} relative to the most stable conformer.

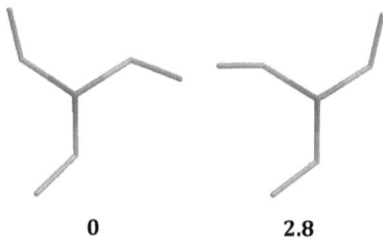

Figure 8: The different conformers of [S$_{2,2,2}$]$^+$, calculated at BP86/TZVP level and sorted by ascending energy. Hydrogen atoms omitted for clarity. Subtitled: the gas phase energies in kJ mol^{-1} relative to the most stable conformer.

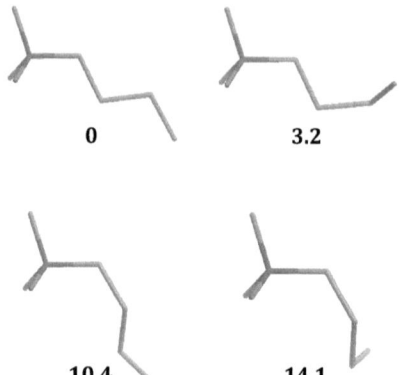

Figure 9: The different conformers of [N$_{1,1,1,4}$]$^+$, calculated at BP86/TZVP level and sorted by ascending energy. Hydrogen atoms omitted for clarity. Subtitled: the gas phase energies in kJ mol^{-1} relative to the most stable conformer.

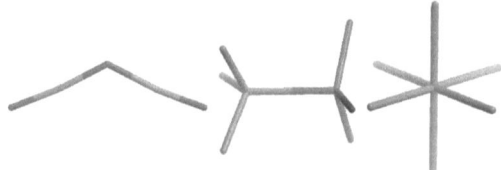

Figure 10: [Dca]$^-$, [OTf]$^-$, and [PF$_6$]$^-$, calculated at BP86/TZVP level.

Table 3: Number of conformers (#) and their gas phase energy range (in kJ mol⁻¹, calculated with BP86/TZVP) for the investigated ions. N.b. for the [Fap]⁻ ion, both *mer* and *fac* stereoisomers were taken into consideration.

ion	#	energy range
[C$_4$MIm]$^+$	4	3.5
[C$_4$MPyr]$^+$	13	15.3
[C$_4$Py]$^+$	4	4.5
[Dca]$^-$	1	-/-
[Fap]$^-$	9	52.0
[N$_{1,1,1,4}$]$^+$	4	14.1
[NTf$_2$]$^-$	2	0.1
[OTf]$^-$	1	-/-
[PF$_6$]$^-$	1	-/-
[S$_{2,2,2}$]$^+$	2	2.8

The results collected in Table 4 to Table 6 and the scatter plot in Figure 11 show that the interaction enthalpies do differ; the maximum absolute relative error of the enthalpies of the most stable conformer against the entire conformer space is 19.0% for H_{HB}, 1.5% for H_{MF}, and 2.2% for H_{vdW}. However, the average correlation coefficient r^2 for all interaction enthalpies, computed with and without the conformer space for all combinations of cations and anions, is 0.9984±0.0016, meaning that there is a very strong linear dependence. This and the multiplied computational time for optimizing all conformers structures – which exponentially rises with the number of rotatable bonds – would make the use of conformers in almost all cases superfluous.

Table 4: COSMO-RS standard hydrogen bonding enthalpies (given in kJ mol⁻¹).

IL	cation[a]	anion[a]	cation[b]	anion[b]
[C$_4$MIm][Dca]	-9.167	-9.154	-9.188	-9.175
[C$_4$MIm][Fap]	-0.287	-0.289	-0.270	-0.272
[C$_4$MIm][NTf$_2$]	-2.783	-2.791	-3.355	-3.360
[C$_4$MIm][OTf]	-6.895	-6.887	-7.025	-7.019
[C$_4$MIm][PF$_6$]	-0.900	-0.911	-0.917	-0.928
[C$_4$MPyr][Dca]	-3.118	-3.077	-3.264	-3.231
[C$_4$MPyr][Fap]	-0.096	-0.095	-0.092	-0.091
[C$_4$MPyr][NTf$_2$]	-0.996	-0.989	-1.212	-1.204
[C$_4$MPyr][OTf]	-2.439	-2.412	-2.549	-2.525
[C$_4$MPyr][PF$_6$]	-0.367	-0.368	-0.382	-0.383
[C$_4$Py][Dca]	-8.283	-8.259	-8.337	-8.317
[C$_4$Py][Fap]	-0.257	-0.258	-0.239	-0.240
[C$_4$Py][NTf$_2$]	-2.589	-2.592	-3.074	-3.074
[C$_4$Py][OTf]	-6.403	-6.389	-6.444	-6.432
[C$_4$Py][PF$_6$]	-0.870	-0.879	-0.876	-0.885
[N$_{1,1,1,4}$][Dca]	-5.189	-5.144	-5.282	-5.230
[N$_{1,1,1,4}$][Fap]	-0.165	-0.164	-0.154	-0.154

[N₁,₁,₁,₄][NTf₂]	-1.697	-1.691	-2.015	-2.004
[N₁,₁,₁,₄][OTf]	-4.099	-4.069	-4.172	-4.138
[N₁,₁,₁,₄][PF₆]	-0.625	-0.628	-0.636	-0.638
[S₂,₂,₂][Dca]	-4.217	-4.186	-4.521	-4.499
[S₂,₂,₂][Fap]	-0.129	-0.129	-0.124	-0.124
[S₂,₂,₂][NTf₂]	-1.331	-1.327	-1.640	-1.637
[S₂,₂,₂][OTf]	-3.269	-3.249	-3.477	-3.463
[S₂,₂,₂][PF₆]	-0.485	-0.488	-0.507	-0.511

[a] Calculated for the most stable conformer only. [b] Calculated using the entire conformer space.

Table 5: COSMO-RS standard misfit enthalpies (given in kJ mol^{-1}).

IL	cation[a]	anion[a]	cation[b]	anion[b]
[C₄MIm][Dca]	21.099	8.045	20.952	7.974
[C₄MIm][Fap]	13.451	18.178	13.314	18.264
[C₄MIm][NTf₂]	14.593	13.210	14.661	13.102
[C₄MIm][OTf]	18.719	9.272	18.563	9.217
[C₄MIm][PF₆]	16.641	6.414	16.466	6.344
[C₄MPyr][Dca]	24.883	9.090	24.746	9.012
[C₄MPyr][Fap]	13.291	17.026	13.169	17.086
[C₄MPyr][NTf₂]	15.815	12.972	16.065	13.002
[C₄MPyr][OTf]	21.725	9.986	21.583	9.923
[C₄MPyr][PF₆]	18.199	5.765	18.060	5.712
[C₄Py][Dca]	20.615	7.885	20.525	7.839
[C₄Py][Fap]	13.121	18.403	13.000	18.477
[C₄Py][NTf₂]	14.185	13.267	14.288	13.158
[C₄Py][OTf]	18.202	9.127	18.094	9.092
[C₄Py][PF₆]	16.067	6.256	15.923	6.192
[N₁,₁,₁,₄][Dca]	21.861	7.855	21.782	7.814
[N₁,₁,₁,₄][Fap]	12.407	18.428	12.328	18.488
[N₁,₁,₁,₄][NTf₂]	13.816	13.052	14.056	12.995
[N₁,₁,₁,₄][OTf]	18.845	9.090	18.755	9.059
[N₁,₁,₁,₄][PF₆]	15.275	5.183	15.176	5.142
[S₂,₂,₂][Dca]	23.768	8.164	23.523	8.090
[S₂,₂,₂][Fap]	11.375	17.355	11.351	17.448
[S₂,₂,₂][NTf₂]	13.827	12.688	14.145	12.722
[S₂,₂,₂][OTf]	20.138	9.201	19.962	9.162
[S₂,₂,₂][PF₆]	16.470	4.589	16.376	4.613

[a] Calculated for the most stable conformer only. [b] Calculated using the entire conformer space.

Table 6: COSMO-RS standard van-der-Waals enthalpies (given in kJ mol^{-1}).

IL	cation[a]	anion[a]	cation[b]	anion[b]
[C₄MIm][Dca]	-40.036	-17.618	-39.508	-17.622
[C₄MIm][Fap]	-33.935	-38.741	-33.449	-38.756
[C₄MIm][NTf₂]	-37.766	-34.396	-37.319	-34.220
[C₄MIm][OTf]	-39.707	-22.150	-39.183	-22.153
[C₄MIm][PF₆]	-36.401	-16.647	-35.847	-16.638
[C₄MPyr][Dca]	-39.878	-17.413	-39.518	-17.422
[C₄MPyr][Fap]	-32.699	-37.348	-32.409	-37.401
[C₄MPyr][NTf₂]	-37.056	-33.599	-36.780	-33.435
[C₄MPyr][OTf]	-39.043	-21.427	-38.696	-21.446
[C₄MPyr][PF₆]	-35.441	-15.587	-35.104	-15.611
[C₄Py][Dca]	-39.129	-17.729	-38.730	-17.734
[C₄Py][Fap]	-33.286	-39.237	-32.919	-39.273
[C₄Py][NTf₂]	-36.901	-34.669	-36.574	-34.502
[C₄Py][OTf]	-38.871	-22.403	-38.475	-22.413

[C₄Py][PF₆]	-35.672	-17.045	-35.257	-17.056
[N₁,₁,₁,₄][Dca]	-35.522	-17.519	-35.279	-17.522
[N₁,₁,₁,₄][Fap]	-28.976	-37.695	-28.768	-37.711
[N₁,₁,₁,₄][NTf₂]	-33.048	-33.827	-32.866	-33.646
[N₁,₁,₁,₄][OTf]	-34.823	-21.619	-34.584	-21.621
[N₁,₁,₁,₄][PF₆]	-31.202	-15.824	-30.963	-15.825
[S₂,₂,₂][Dca]	-36.165	-18.308	-35.930	-18.314
[S₂,₂,₂][Fap]	-29.629	-39.386	-29.482	-39.367
[S₂,₂,₂][NTf₂]	-33.558	-35.191	-33.427	-34.977
[S₂,₂,₂][OTf]	-35.431	-22.683	-35.212	-22.664
[S₂,₂,₂][PF₆]	-31.676	-16.984	-31.777	-16.996

a Calculated for the most stable conformer only. b Calculated using the entire conformer space.

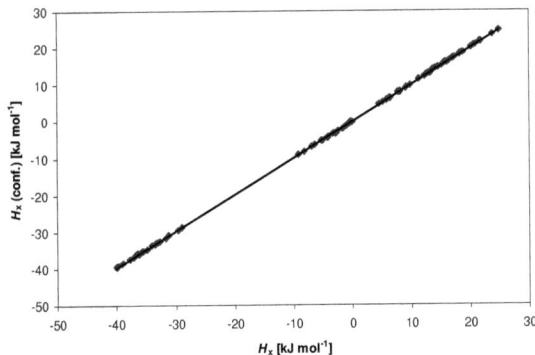

Figure 11: The correlation of all calculated interaction enthalpies H_x (x = HB, vdW, MF) for an IL consisting only of the two global minum ions and H_x (conf.), where the entire conformer space was taken into consideration.

2.2. Molecular Structures

2.2.1. Calculated Weakly Coordinating Anions and Related Species

In the following, calculated structures are given for the basic types of WCAs, Lewis acids and their fluoride adducts. Variations arise from the exchange of ligands, including the exchange of fluorine bound to the central atom by doubly-bound oxygen if applicable.

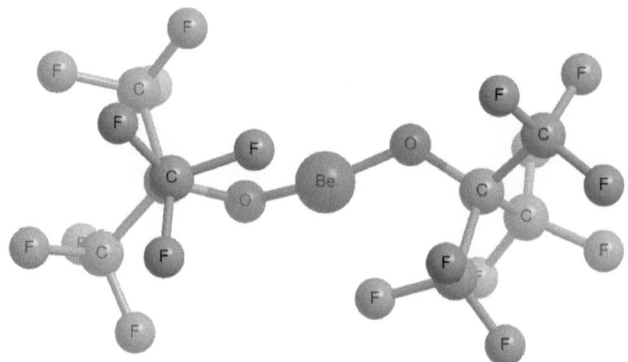

Figure 12: Structure of Be(pftb)$_2$, calculated at the BP86/SV(P) level.

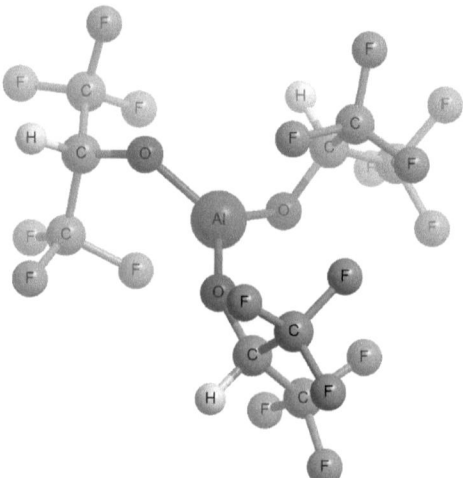

Figure 13: Structure of Al(hfip)$_3$, calculated at the BP86/SV(P) level.

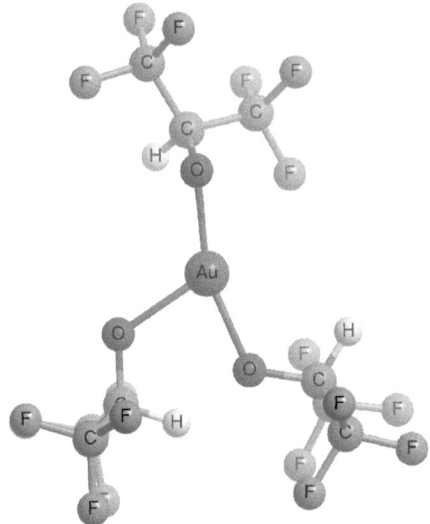

Figure 14: Structure of Au(hfip)$_3$, calculated at the BP86/SV(P) level.

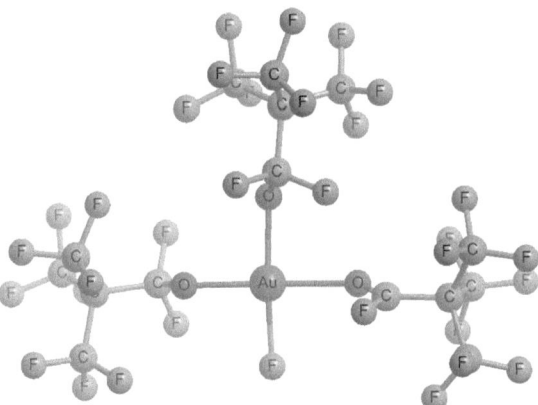

Figure 15: Structure of Au(pfn^5)$_3$ (with one cleaved ligand, resulting in a fluoroacyl complex; see text), calculated at the BP86/SV(P) level.

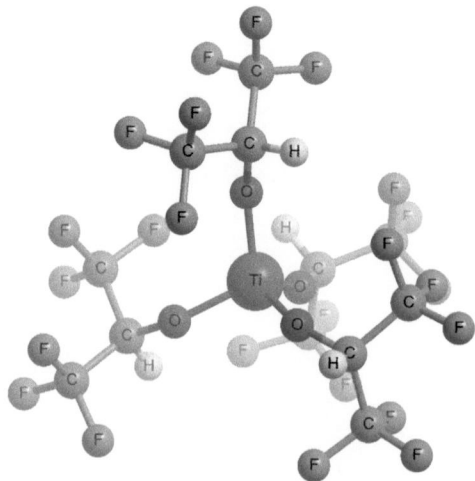

Figure 16: Structure of Ti(hfip)$_4$, calculated at the BP86/SV(P) level.

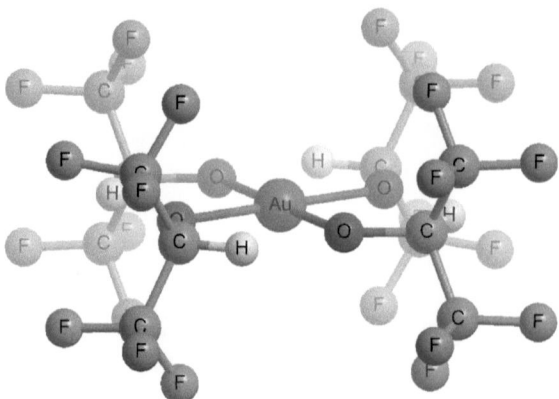

Figure 17: Structure of [Au(hfip)$_4$]$^-$, calculated at the BP86/SV(P) level.

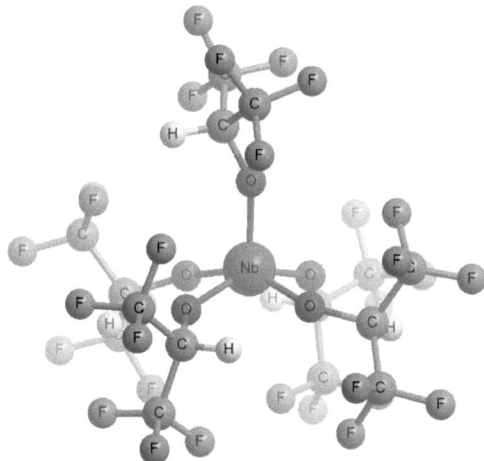

Figure 18: Structure of Nb(hfip)$_5$, calculated at the BP86/SV(P) level.

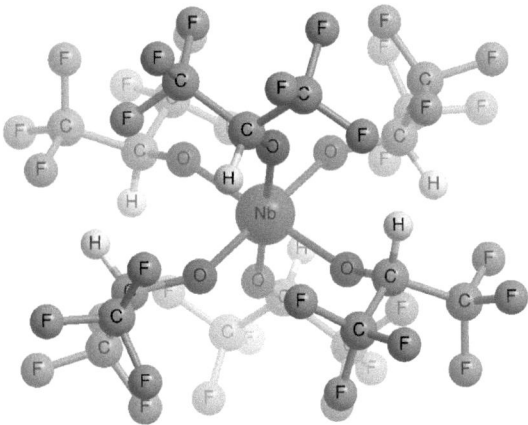

Figure 19: Structure of [Nb(hfip)$_6$]$^-$, calculated at the BP86/SV(P) level.

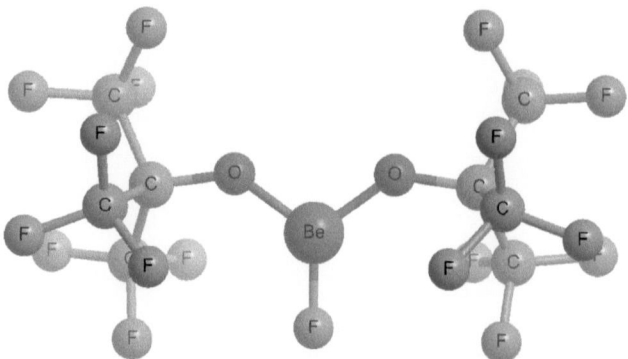

Figure 20: Structure of [Be(pftb)$_2$F]$^-$, calculated at the BP86/SV(P) level.

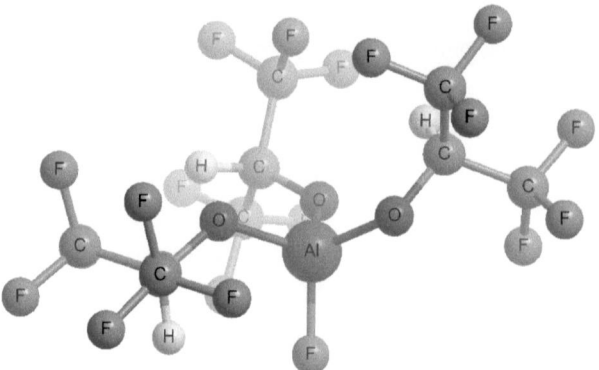

Figure 21: Structure of [Al(hfip)$_3$F]$^-$, calculated at the BP86/SV(P) level.

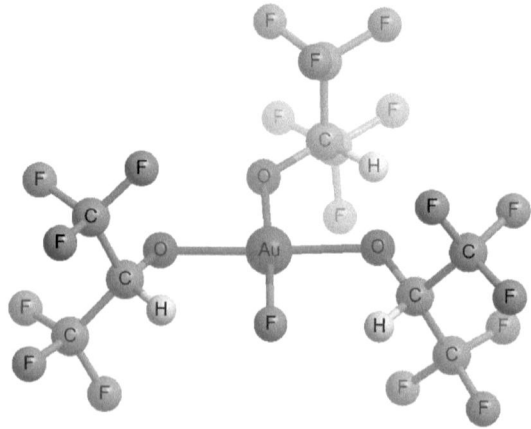

Figure 22: Structure of [Au(hfip)$_3$F]$^-$, calculated at the BP86/SV(P) level.

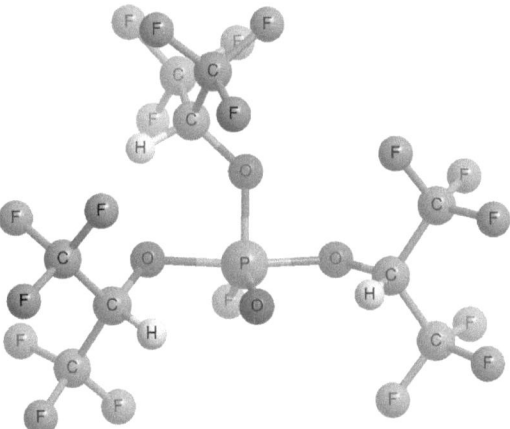

Figure 23: Structure of [P(hfip)$_3$OF]$^-$, calculated at the BP86/SV(P) level. If the F atom and one hfip ligand are made to stand in axial position as VSEPR theory would lead to expect, the resulting structure is 8.3 kJ mol^{-1} less stable in the gas phase (BP86/SV(P)) than the shown one, probably due to sterical crowding.

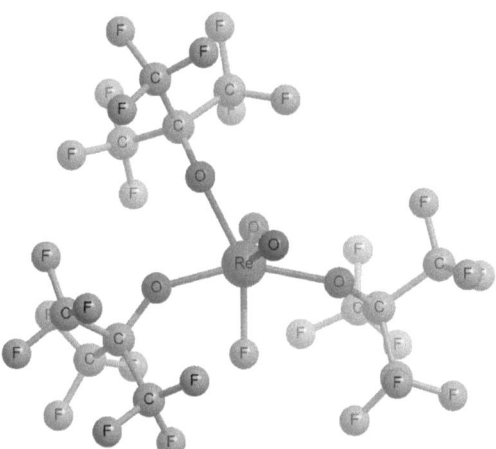

Figure 24: Structure of [Re(pftb)$_3$O$_2$F]$^-$, calculated at the BP86/SV(P) level.

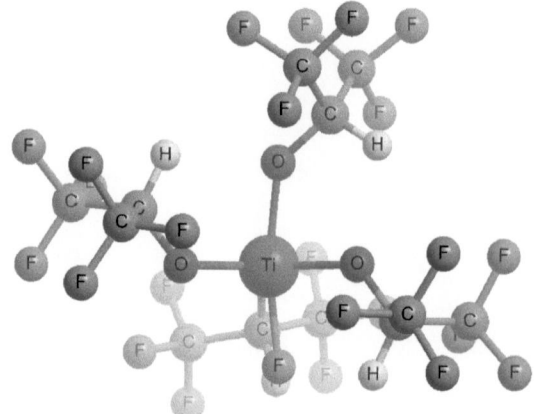

Figure 25: Structure of [Ti(hfip)$_4$F]$^-$, calculated at the BP86/SV(P) level.

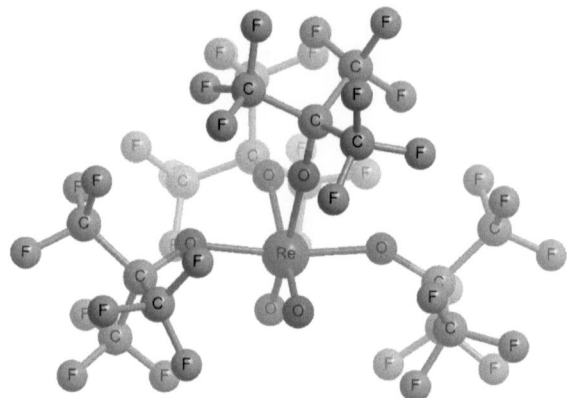

Figure 26: Structure of [Re(pftb)$_4$O$_2$]$^-$, calculated at the BP86/SV(P) level.

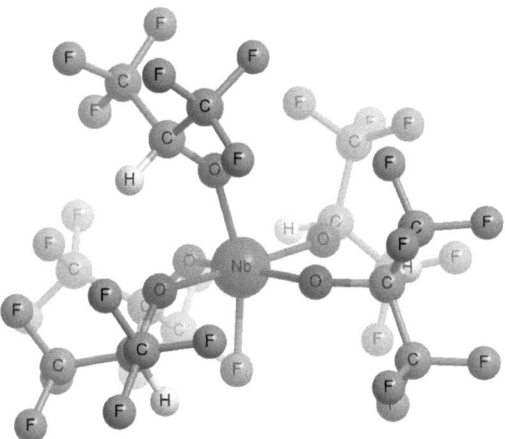

Figure 27: Structure of [Nb(hfip)$_5$F]$^-$, calculated at the BP86/SV(P) level.

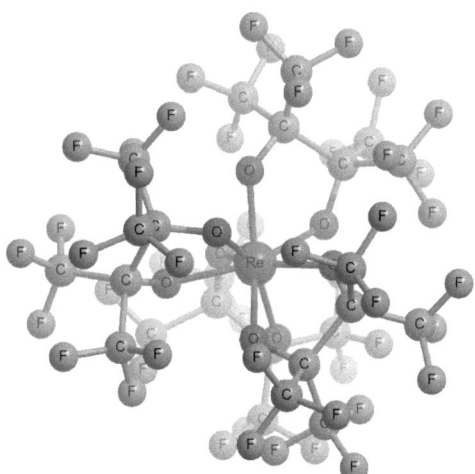

Figure 28: Structure of [Re(pfp)$_4$]$^-$, calculated at the BP86/SV(P) level.

The following tables give an overview to the structural parameters for all calculated species, including the angle of the central atom and the ligands, the distance between central atom and ligand oxygen, the distance between this oxygen and the first carbon atom of the ligand, and, if applicable, the distance between central atom and directly bound fluorine or doubly-bound oxygen. If there are close contacts between ligand-bound fluorine atoms and the central atoms, the shortest one is given. A deviation of 2 pm or 2° was not explicitly stated, as it is the usual precision limit of DFT methods.

Table 7: WCAs of the type [ML₄]⁻ or [ML₂]⁻ for L = pfp, respectively.

M	L	∠M-O-C [°]	d(M-O) [pm]	d(O-C) [pm]	d(M=O) [pm]	structure
ReO₂	pftb	153±16	204±4	136	174	Figure 26
Fe	hfip	128	185	137		Figure 16
Cu	hfip	122	189	137		Figure 17
Ag	pfp	113	202	139		Figure 17
Ag	hfip	119	204	138		Figure 17
Au	pfp	113	204	139		Figure 17
Au	dfc[6]	119	205	138		Figure 17
Au	hfip	119	205	138		Figure 17
Au	bpfipm	120	207	137		Figure 17
Au	pfn[5]	123	207	132		Figure 17
B	pfp	111	149	138		Figure 16
B	hfip	123	148	138		Figure 16
Al	pfp	113	179	138		Figure 16
Al	hfip	134	177	137		Figure 16
Al	dfc[6]	135	177	136		Figure 16
Al	pfn[5]	135	178	131		Figure 16
Al	mpfc[6]	139	178	137		Figure 16
Al	hftb	140	178	137		Figure 16
Al	pfbc[7]	144	177	134		Figure 16
Al	pfbc[8]	148	177	134		Figure 16
Al	pfad	149	177	134		Figure 16
Al	bpfipm	149	178	135		Figure 16
Al	pftb	150	177	135		Figure 16
Ga	hfip	122	186	138		Figure 16
Ga	hftb	132	187	138		Figure 16
Ga	pftb	140	186	136		Figure 16
PO	pfp	117	177±5	138	150	Figure 25
PO	hfip	124±6	175±7	139	152	Figure 25
PO	hftb	130±3	176±8	140	151	Figure 25
PO	pftb	137±10	176±7	138	151	Figure 25

Table 8: WCAs of the type [ML₆]⁻ or [ML₃]⁻ for L = pfp, respectively. Structures are similar to Figure 19.

M	L	∠M-O-C [°]	d(M-O) [pm]	d(O-C) [pm]
Nb	pfp	124	202	138
Nb	pfn[5]	141	202	134
Nb	hfip	144	200	137
Nb	dfc[6]	146	199	137
Nb	hftb	160±6	202	138
Nb	pftb	166	205	138
Ta	hfip	146	200	137
Ta	dfc[6]	147	200	137
Ta	hftb	161±6	202	138
Ta	pftb	166	205	138
Au	pfp	116	209	139
Au	hftb	134	218±3	139
P	pfp	119	174	138
P	hfip	127	176	140
P	hftb	142±4	181±4	142
As	pfp	118	187	138
Sb	pfp	117	207	138
Sb	pfn[5]	128	208	134
Sb	hfip	129	205	139

Sb	dfc[6]	131	205	139
Sb	hftb	138	210	141
Sb	pftb	146	216	139
Bi	hfip	127	219	138

Table 9: WCAs of the type [ML$_3$]$^-$. Structures are similar to Figure 13.

M	L	∠ M-O-C [°]	d(M-O) [pm]	d(O-C) [pm]	d(M-F) [pm]
Be	mpfc[6]	135	155	136	
Be	pftb	141	155	135	
Be	pfad	146	154	133	
Be	bpfipm	134±6	155	136	
Ni	bpfipm	128±11	183	136	
Zn	bpfipm	125	190	136	
Zn	pftb	128	191	135	263
Zn	pfad	133	189	134	

Table 10: WCAs of the type [ML$_5$]$^-$. Structures are similar to Figure 18.

M	L	∠ M-O-C [°]	d(M-O) [pm]	d(O-C) [pm]
Ti	hfip	133	190±3	138
Ti	hftb	150	191±4	138
Ti	pftb	161±6	192±4	136
Zr	hfip	141±10	204±3	137
Zr	pftb	167±5	205±3	136
Hf	hfip	138±4	204	137
Hf	pftb	170±3	205±3	136
Si	hfip	130	175±3	138

Table 11: Lewis acids of the type ML$_2$. Structures are similar to Figure 12.

M	L	∠ M-O-C [°]	d(M-O) [pm]	d(O-C) [pm]	d(M-F) [pm]
Be	pftb	125	148	137	209
Be	mpfc[6]	127	147	139	243
Be	bpfipm	136±14	145	138	223
Be	pfad	137±5	144	136	
Ni	bpfipm	121±7	182	137	201
Zn	pfad	114	184	136	250
Zn	pftb	120	183	137	241
Zn	bpfipm	127±3	181	138	246

Table 12: Lewis acids of the type ML$_3$.

M	L	∠ M-O-C [°]	d(M-O) [pm]	d(O-C) [pm]	d(M-F) [pm]	d(M=O) [pm]	structure
ReO$_2$	pftb	146	195	138		173	Figure 23
Fe	hfip	126	178	139	262		Figure 13
Cu	hfip	120±3	185	137	215		Figure 14
Ag	hfip	119	203	137	243		Figure 14
Au [a]	pfn[5]	115	203±3	133	196		Figure 15[a]
Au	bpfipm	118±5	203	139	235		Figure 14
Au	dfc[6]	122±3	203±5	138			Figure 14
Au	hfip	122	204±6	138			Figure 14
B	hfip	128±4	137	140			Figure 13
Al	dfc[6]	125±11	174	138	228		Figure 13
Al	bpfipm	130±8	174	138	213		Figure 13
Al	hfip	132±16	174	138	215		Figure 13
Al	pfn[5]	132±8	175	133	214		Figure 13

M	L	∠M-O-C [°]	d(M-O) [pm]	d(O-C) [pm]			Figure
Al	hftb	133±12	173	139	206		Figure 13
Al	pftb	134±15	174	137	212		Figure 13
Al	pfad	137±16	173±3	137	219		Figure 13
Al	pfbc[7]	137±17	173±3	136	211		Figure 13
Al	mpfc[6]	139±19	172±3	139	228		Figure 13
Al	pfbc[8]	139±17	173±3	136	214		Figure 13
Ga	hfip	125±7	182	139	240		Figure 13
Ga	pftb	127±5	183	138	234		Figure 13
Ga	hftb	128±4	182	140	236		Figure 13
PO	hfip	121	163	142		149	Figure 21
PO	hftb	132±5	163	144		149	Figure 21
PO	pftb	135	164	141		148	Figure 21

[a] One ligand decomposes, forming a fluoroacyl adduct with d(Au-O) = 217 pm.

Table 13: Lewis acids of the type ML_4. Structures are similar to Figure 16.

M	L	∠M-O-C [°]	d(M-O) [pm]	d(O-C) [pm]
Ti	hfip	151±8	181	139
Ti	hftb	161±9	180	139
Ti	pftb	165	180	138
Zr	pftb	168	196	137
Zr	hfip	172±3	196	138
Hf	pftb	165	196	137
Hf	hfip	166±8	196	138
Si	hfip	129	166	141

Table 14: Fluoride adducts of the type $[ML_2F]^-$. Structures are similar to Figure 20.

M	L	∠M-O-C [°]	d(M-O) [pm]	d(O-C) [pm]	d(M-F) [pm]
Be	bpfipm	119	156	136	148
Be	mpfc[6]	129	156	136	149
Be	pftb	137	157	134	146
Be	pfad	141	157	133	145
Ni	bpfipm	119	182	136	177
Zn	bpfipm	113	190	137	184
Zn	pfad	127	192	133	180
Zn	pftb	127	192	134	180

Table 15: Lewis acids of the type ML_5. Structures are similar to Figure 18.

M	L	∠M-O-C [°]	d(M-O) [pm]	d(O-C) [pm]
Nb	hfip	142±6	194±3	139
Nb	pfn[5]	144±3	195±3	137
Nb	dfc[6]	152±17	194	138
Nb	hftb	159±4	194±3	140
Nb	pftb	174	196±3	139
Ta	hfip	143±5	195±3	139
Ta	dfc[6]	152±17	194	138
Ta	hftb	159±4	195±3	140
Ta	pftb	174	196±3	139
Au	hftb	134±5	213±4	138
P	hfip	130	167±3	141
P	hftb	139±5	170±4	144
Sb	hfip	125	200	141
Sb	pfn[5]	125	203	136
Sb	dfc[6]	128	201	140
Sb	hftb	136	203	142

Sb	pftb	140±3	207±3	140	
Bi	hfip	123	214	140	

Table 16: Fluoride adducts of the type [ML$_3$F]$^-$.

M	L	∠ M-O-C [°]	d(M-O) [pm]	d(O-C) [pm]	d(M-F) [pm]	d(M=O) [pm]	structure
ReO$_2$	pftb	139±3	207±6	137	192	175	Figure 24
Fe	hfip	122	185	138	180		Figure 21
Cu	hfip	120	188	137	182		Figure 22
Ag	hfip	117	203	138	199		Figure 22
Au	hfip	117	204	138	201		Figure 22
Au	dfc^6	117	205	138	200		Figure 22
Au	bpfipm	119	205	138	200		Figure 22
Au	pfn^5	122	206	132	196		Figure 22
B	hfip	121	148	138	141		Figure 21
Al	pfad	119	178	134	169		Figure 21
Al	hfip	126	170	137	170		Figure 21
Al	dfc^6	129±6	178	136	170		Figure 21
Al	bpfipm	131±7	179	137	170		Figure 21
Al	mpfc^6	136	178	137	170		Figure 21
Al	pfn^5	137±4	179	130	168		Figure 21
Al	hftb	138	178	137	170		Figure 21
Al	pfbc^7	140	178	134	169		Figure 21
Al	pfbc^8	144	178	134	169		Figure 21
Al	pftb	147±4	178	135	169		Figure 21
Ga	hfip	120	187	138	179		Figure 21
Ga	pftb	136	187	136	178		Figure 21
Ga	hftb	138	186	137	178		Figure 21
PO	hfip	123±5	176±7	139	163	152	Figure 23
PO	hftb	129±4	177±6	140	164	151	Figure 23
PO	pftb	133	176±5	138	169	150	Figure 23

Table 17: Fluoride adducts of the type [ML$_5$F]$^-$. Structures are similar to Figure 27.

M	L	∠ M-O-C [°]	d(M-O) [pm]	d(O-C) [pm]	d(M-F) [pm]
Nb	hfip	136	200±3	138	195
Nb	pfn^5	144±4	202	133	191
Nb	dfc^6	146±7	199	136	194
Nb	hftb	160±9	200	138	194
Nb	pftb	163±7	202	137	191
Ta	hfip	138	200	137	196
Ta	dfc^6	147±6	200	136	194
Ta	hftb	160±9	201	138	195
Ta	pftb	166±9	202	136	192
Au	hftb	135±4	215±5	138	200
P	hfip	127	174	139	168
P	hftb	139±4	177	141	165
Sb	hfip	122	204	139	199
Sb	pfn^5	127	208	134	194
Sb	dfc^6	129	205	139	196
Sb	hftb	136	207	140	197
Sb	pftb	143±3	209	138	195
Bi	hfip	120	217	139	208

Table 18: Fluoride adducts of the type [ML$_4$F]$^-$. Structures are similar to Figure 25.

M	L	∠ M-O-C [°]	d(M-O) [pm]	d(O-C) [pm]	d(M-F) [pm]
Ti	hfip	132±3	189±3	138	185
Ti	hftb	145±5	190±4	138	183
Ti	pftb	152±7	190±4	136	180
Zr	hfip	136	204	137	200
Zr	pftb	162±10	204±3	135	196
Hf	hfip	136	204	137	199
Hf	pftb	164±9	204±3	145	196
Si	hfip	127	175±3	138	167

For [Re(pfp)$_4$]$^-$ (Figure 28), ∠ M-O-C = 126±3°, d(M-O) = 207±6 pm, d(O-C) = 138 pm. Especially in WCAs, sterical crowding usually precludes the formation of bonds between central atom and ligand-bound fluorine atoms. The M-O distance as well as the M-O-C angle give information about the ionicity of the ligand-metal bond. A high value (up to 180° for the angle) means a completely ionic bond; a low value (down to 109° for the angle) a completely covalent one. Both values are of course highly influenced by sterical hindrance. The most extreme angles in WCAs are for [M(pftb)$_5$]$^-$ with M = Ti, Zr, Hf (161...170°) and for [ML$_6$]$^-$ with M = Nb, Ta and L = hftb, pftb (160...166°). Here, empty d orbitals, which support a linear coordination, influence the geometry. Furthermore, the strong metallic character of the central atoms and the sterical repulsion of the ligands cause an almost completely ionic character of the metal-ligand bond. Likewise, the ligand affinity in these compounds is quite low (109...254 kJ mol^{-1}, see Table 29, Table 31 and Table 34). The M-O-C angle is smallest (111...113°) and very close to the tetrahedral angle in the sterically little hindered pfp chelate WCAs, i.e. [M(pfp)$_2$]$^-$, where M = B, Ag, Au. Incidentally, a [Cu(pfp)$_2$]$^-$ WCA cannot form because of the Cu^{3+} ion being much smaller than Ag^{3+} or Au^{3+}, thereby increasing the ring strain; a geometry optimization with a forced bicyclic starting geometry of this compound results in C-C cleavage, forming a [Cu(pfp)]$^-$ unit with two additionally coordinated (CF$_3$)$_2$CO molecules.

The O-C bond length is usually 137 or 138 pm in most WCAs. Exceptions with 140 or more pm are found in [P(hfip)$_6$]$^-$, [P(hftb)$_6$]$^-$, [P(hftb)$_4$O]$^-$, and [Sb(hftb)$_6$]$^-$. Here, the ligand affinity is also quite small (49...266 kJ mol^{-1}, see Table 29 and Table 34), meaning that the ligands, due to sterical overcrowding and relatively low degree of fluorination, cannot form a bond to the oxygen atom that is as strong as in other WCAs. These system could also be described with the resonance structures [M(OR)$_n$]$^-$ and R$^+$ [O=M(OR)$_{n-1}$]$^{2-}$, where OR is a single ligand. Especially for M = P, the formation of a P=O double bond has significant weight, allowing sterical tension to be alleviated.

Table 19 to Table 21 show available crystal data from literature for some of the aforementioned species except for [M(CF$_3$)$_4$]$^-$ treated in section 3.2 (M = Cu, Ag, Au). Coordination of a countercation may lead to distortion of the anion, as the experimental values in Table 19 show. The smaller angle as well as the larger M-O distance belong to the ligands to which the countercation coordinates; these angles and bond lengths are typically also the values with the greatest accordance with the calculated ones (see Table 7 and Table 8).

For the B-CF$_3$ species in Table 20 and also the Lewis acids shown in Table 21, the agreement with the calculated gas phase values (see Table 12 and Table 13) is quite excellent except for the M-O-C angle in Ti(hfip)$_4$ which is decreased by 10° in the lattice structure due to the additional coordination of two solvent molecules forcing an octa- instead of a tetrahedral geometry.

Table 19: Experimental data for WCAs of the type [ML$_n$]$^-$.

compound		∠ M-O-C [°]	d(M-O) [pm]	d(O-C) [pm]	ref.
Cs[Al(hfip)$_4$]	exp.	137±1	173±2	131±6	19
[Ag(C$_2$H$_4$Cl$_2$)$_2$] [Al(hfip)$_4$]	exp.	130±6	174±2	138±1	20
[Al(hfip)$_4$]$^-$	calc.	134	177	137	
[Ag(CH$_2$Cl$_2$)][Al(hftb)$_4$]	exp.	138, 143±1	171, 177	139±1	20
[Al(hftb)$_4$]$^-$	calc.	140	178	137	
Cs[Al(pftb)$_4$]	exp.	153±1, 163±2	171±2	131±4	19
[Ag(C$_2$H$_4$Cl$_2$)$_3$][Al(pftb)$_4$]	exp.	150±2	172±1	134±1	20
[Al(pftb)$_4$]$^-$	calc.	150	177	135	
[Ag(C$_6$H$_5$F)][Nb(hfip)$_6$]	exp.	137±3, 157±4	190±1, 201±2	139±1	a
[Li$_2$(1,2-C$_6$H$_4$F$_2$)][Nb(hfip)$_6$]$_2$	exp.	134±2, 147±1, 165±1	191±2, 206±1	139±1	a
[Nb(hfip)$_6$]$^-$	calc.	144	200	137	

[a] This study.

Table 20: Experimental and calculated (BP86/SV(P)) data for B-CF$_3$ compounds.

compound		∠ C-B-C [°]	d(B-C) [pm]	d(C-F) [pm]	d(B-F) [pm]	ref.
Cs[B(CF$_3$)$_4$]	exp.	109	162	135		21
[B(CF$_3$)$_4$]$^-$	calc.	109	165	137		
[Co(CO)$_5$][B(CF$_3$)$_3$F]	exp.	110±1	163	135±1	141	22
[B(CF$_3$)$_3$F]$^-$	calc.	111	165	137	143	
B(CF$_3$)$_3$	calc.	120	161	136±1		

Table 21: Experimental data for Lewis acids of the type M(hfip)$_n$.

M	n		∠M-O-C [°]	d(M-O) [pm]	d(O-C) [pm]	d(M-O) [pm]	ref.
Ga[a]	3	exp.	125±2	181±1	139		23
		calc.	125±7	182	139		
P O	3	exp.	121±1	158±1	142	145	24
		calc.	121	163	142	149	
Ti[b]	4	exp.	141±7	184	136±1		25
		calc.	151±8	181	139		
Zr	4	exp.	165±7	192±1	139±2		26
		calc.	172±3	196	138		

[a] One 4-dimethylaminopyridine molecule is additionally coordinated to the Ga center. [b] Two acetonitrile molecules are additionally coordinated to the Ti center.

The geometries for all WCAs, their associated Lewis acids and their fluoride adducts treated in this study as well as all related species needed to calculate the quantities given in section 3.3 are accessible via ssh from the intranet at the University of Freiburg at the IP address 132.230.120.184 in the directory /share/ilgroup/WCAs and its subdirectories which amount to a total size of ≈ 3.6 GB.

2.2.2. Calculated Ion Pairs

In recent times, growing amount of consideration has gone to the question which structures ions and ion pairs assume in the liquid or in the gas phase for elucidation of various thermodynamic properties of ionic liquids.[27,28,29,30,31,32,33,34,35,36,37,38,39,40] In our approach, ion pair structures were determined by calculating partial charges according to Mulliken for cations and anions separately.[41] Up to 10 different conformers were then generated and optimized with B3LYP/6-31G*.[42,43] The one to three conformers with the lowest energy were then optimized with G3MP2[44] using Gaussian 03[45] giving the global minimum (all G3MP2 optimizations were done by Dr. V. N. Emel'yanenko, Institute of Physical Chemistry, University of Rostock). Starting from this, optimizations up to the BP86/TZVP+COSMO level were performed as detailed above. Below are gas phase structures for all ILs for which pairs were calculated. The lengths of directed interactions between anion and cation were plotted as well. Table 22 gives all calculated total energies.

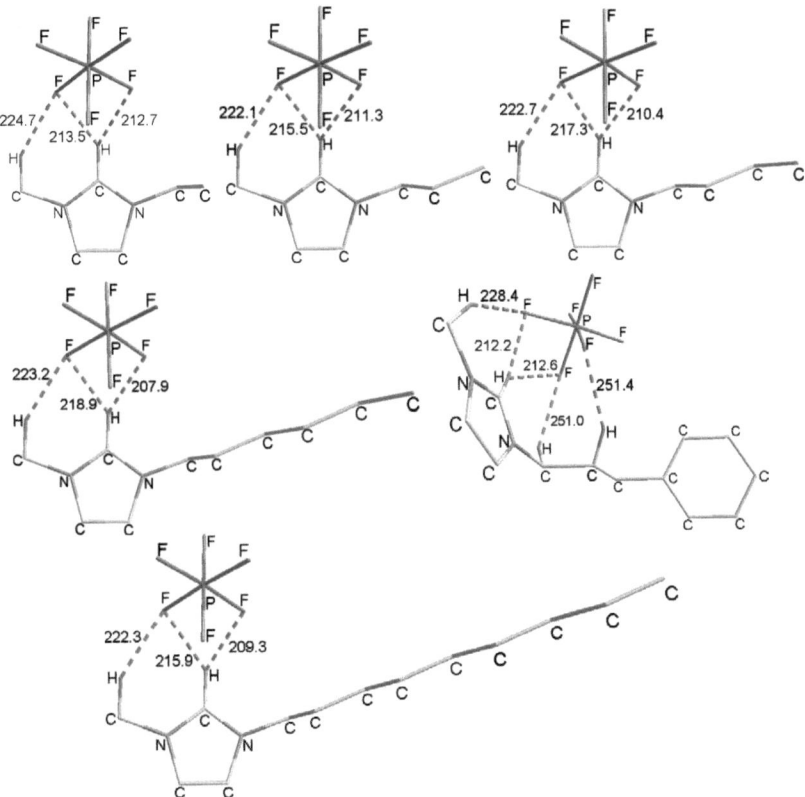

Figure 29: [C$_2$MIm][PF$_6$], [C$_3$MIm][PF$_6$], [C$_4$MIm][PF$_6$], [C$_6$MIm][PF$_6$], [C$_8$MIm][PF$_6$], and [φCCCMIm][PF$_6$], calculated at the BP86/TZVP level. Non-coordinating hydrogen atoms were removed.

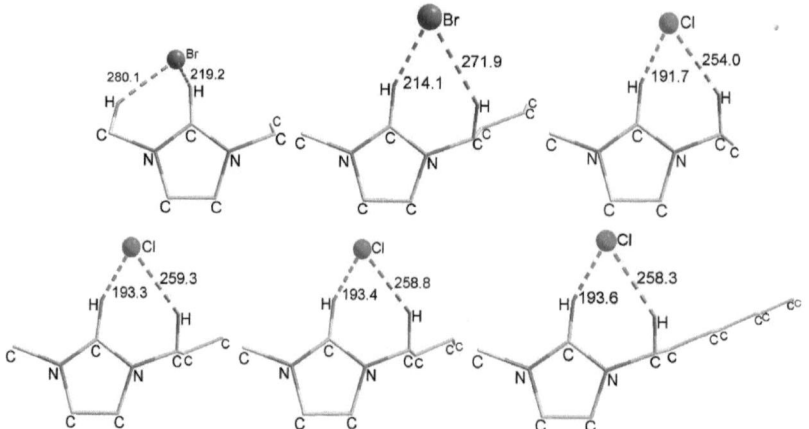

Figure 30: [C₂MIm]Br, [C₄MIm]Br, [C₂MIm]Cl, [C₃MIm]Cl, [C₄MIm]Cl, and [C₈MIm]Cl, calculated at the BP86/TZVP level. Non-coordinating hydrogen atoms were removed.

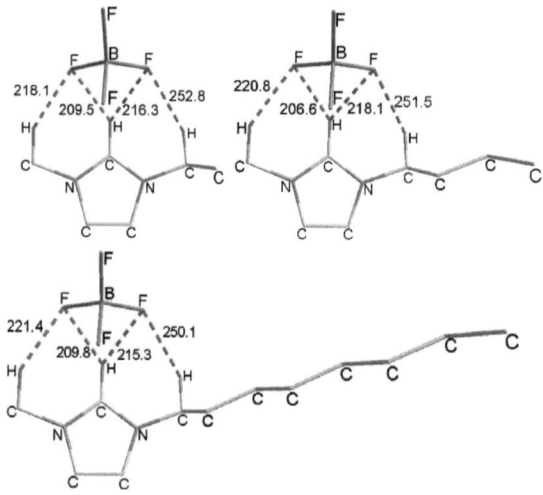

Figure 31: [C₂MIm][BF₄], [C₄MIm][BF₄], and [C₈MIm][BF₄], calculated at the BP86/TZVP level. Non-coordinating hydrogen atoms were removed.

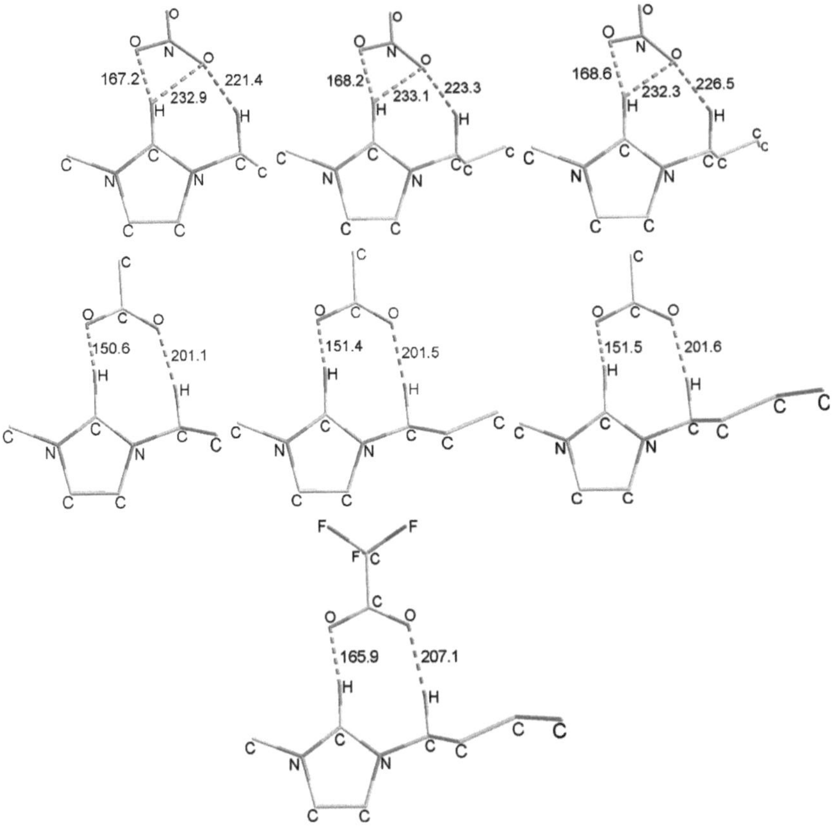

Figure 32: [C$_2$MIm][NO$_3$], [C$_3$MIm][NO$_3$], [C$_4$MIm][NO$_3$], [C$_2$MIm][OAc], [C$_3$MIm][OAc], [C$_4$MIm][OAc], and [C$_4$MIm][OTfa], calculated at the BP86/TZVP level. Non-coordinating hydrogen atoms were removed.

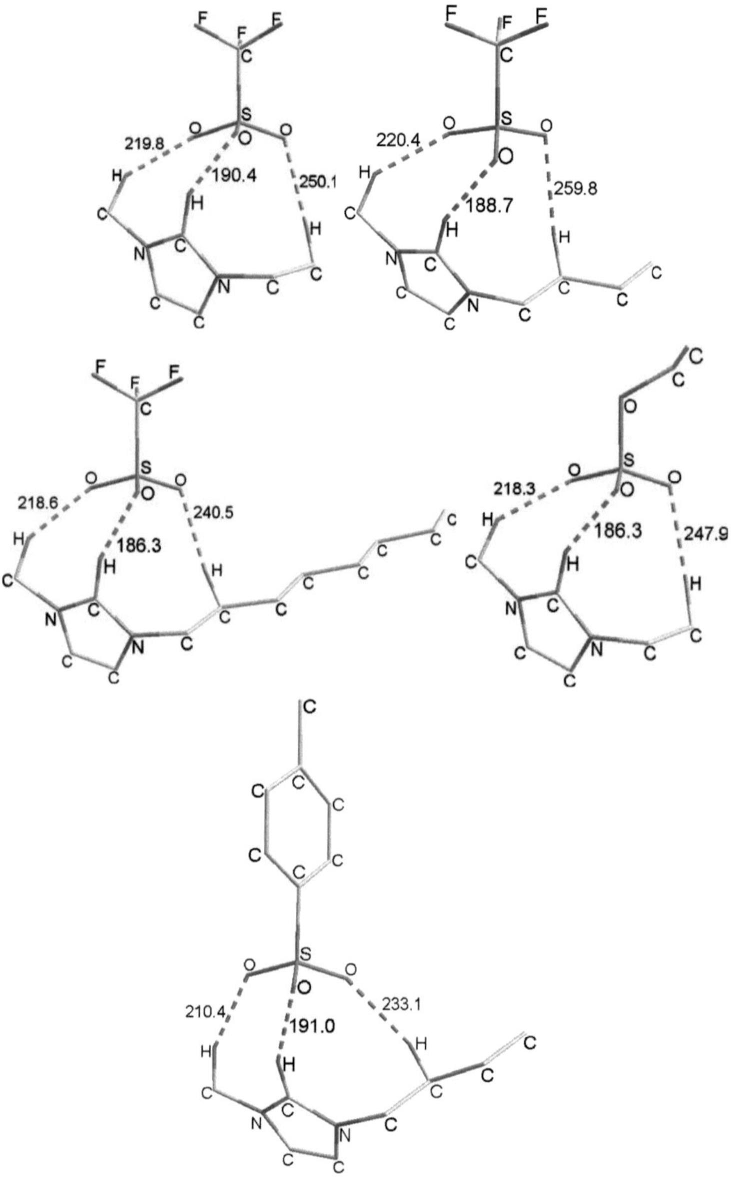

Figure 33: [C$_2$MIm][OTf], [C$_4$MIm][OTf], [C$_8$MIm][OTf], [C$_2$MIm][C$_2$SO$_4$], and [C$_4$MIm][OTos], calculated at the BP86/TZVP level. Non-coordinating hydrogen atoms were removed.

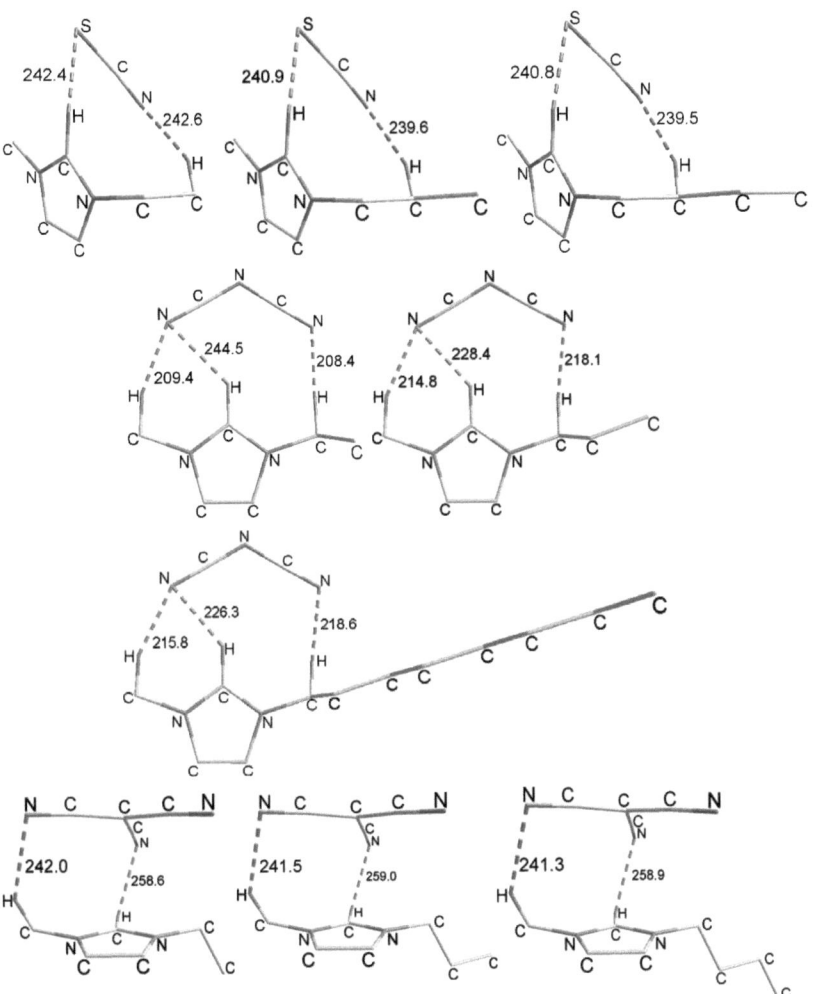

Figure 34: [C₂MIm][SCN], [C₃MIm][SCN], [C₄MIm][SCN], [C₂MIm][Dca], [C₃MIm][Dca], [C₄MIm][Dca], [C₈MIm][Dca], [C₂MIm][Tcm], [C₃MIm][Tcm], and [C₄MIm][Tcm], calculated at the BP86/TZVP level. Non-coordinating hydrogen atoms were removed.

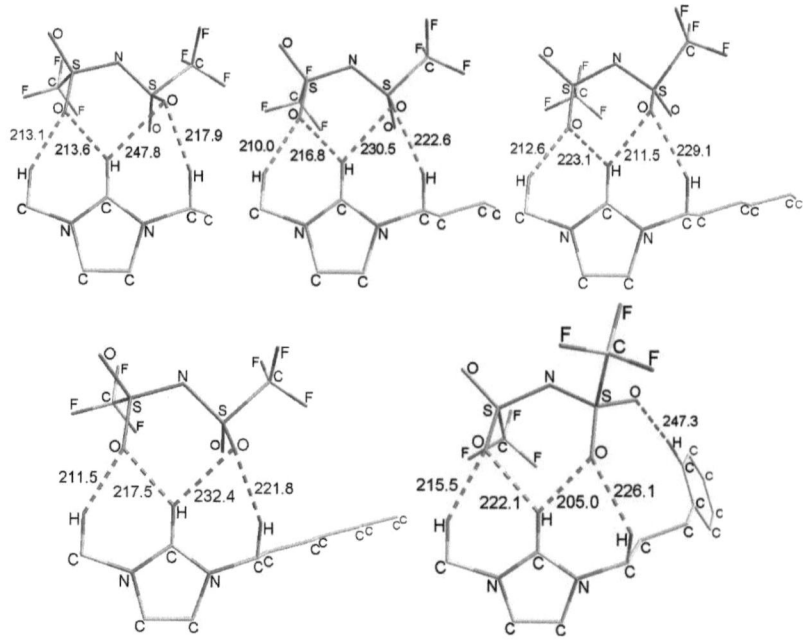

Figure 35: [C$_2$MIm][NTf$_2$], [C$_4$MIm][NTf$_2$], [C$_6$MIm][NTf$_2$], [C$_8$MIm][NTf$_2$], and [φCCCMIm][NTf$_2$], calculated at the BP86/TZVP level. Non-coordinating hydrogen atoms were removed.

Table 22: Total gas phase energies (in hartree at 0 K) of all ion pairs used in this study. Index 1 denotes BP86/TZVP, index 2 G3MP2 values.

IL	U_1	U_2
[C$_2$MIm][BF$_4$]	-769.529287	
[C$_2$MIm][C$_2$SO$_4$]	-1123.372586	-1121.579345
[C$_2$MIm][Dca]	-585.404682	-584.263084
[C$_2$MIm][NO$_3$]	-625.328463	-624.153690
[C$_2$MIm][NTf$_2$]	-2172.637212	-2169.570236
[C$_2$MIm][OAc]	-573.461760	-572.331512
[C$_2$MIm][OTf]	-1306.654636	
[C$_2$MIm][PF$_6$]	-1285.794004	-1283.874044
[C$_2$MIm][SCN]	-836.012268	-834.637357
[C$_2$MIm][Tcm]	-661.621417	-660.350560
[C$_2$MIm]Br	-2919.341016	
[C$_2$MIm]Cl	-805.151033	-803.911215
[C$_3$MIm][Dca]	-624.731871	-623.497256
[C$_3$MIm][NO$_3$]	-664.655306	-663.389082
[C$_3$MIm][OAc]	-612.788376	-611.567220
[C$_3$MIm][PF$_6$]	-1325.120790	-1323.109917
[C$_3$MIm][SCN]	-875.338808	-873.872519
[C$_3$MIm][Tcm]	-700.948794	
[C$_3$MIm]Cl	-844.477842	-843.147092
[C$_4$MIm][BF$_4$]	-848.182299	
[C$_4$MIm][Dca]	-664.058120	-662.731410
[C$_4$MIm][NO$_3$]	-703.981527	-702.623897
[C$_4$MIm][NTf$_2$]	-2251.290158	-2248.043563
[C$_4$MIm][OAc]	-652.114549	-650.801968

[C₄MIm][OTf]	-1385.307800	
[C₄MIm][OTfa]	-949.986774	
[C₄MIm][OTos]	-1318.576904	
[C₄MIm][PF₆]	-1364.447182	-1362.345172
[C₄MIm][SCN]	-914.665094	-913.107211
[C₄MIm][Tcm]	-740.275002	-738.821026
[C₄MIm]Br	-2997.993088	
[C₄MIm]Cl	-883.804064	-882.381865
[C₆MIm][NTf₂]	-2329.943002	
[C₆MIm][PF₆]	-1443.099728	
[C₈MIm][BF₄]	-1005.487112	
[C₈MIm][Dca]	-821.362770	
[C₈MIm][NTf₂]	-2408.594939	
[C₈MIm][OTf]	-1542.612443	
[C₈MIm][PF₆]	-1561.078121	
[C₈MIm]Cl	-1041.108734	
[φCCCMIm][NTf₂]	-2443.093024	
[φCCCMIm][PF₆]	-1556.250374	

2.2.3. Experimental Structures

To further elucidate the binding situation of weakly coordinated systems, some compounds were synthesized and structurally characterized.

Structures of WCAs with the hfip ligand

Li[Nb(hfip)₆] was synthesized from NbCl₅ and 6 equivalents of Li-hfip according to eqn. (VI) in section 3.3.2. Crystallization took place by undercooling a saturated solution of the product in 1,2-difluorobenzene to -30 °C. The structure (Figure 36; space group: C 2/c, R_1 = 3.03%) contains one solvent molecule that bridges two Li[Nb(hfip)₆] moieties, forming an isolated superstructure (Figure 37). There's a rather strong coordination of the Li cation to the – sterically little hindered – O atoms of the anion. As Figure 38 shows, the cation forms the center of a distorted tetragonal pyramid with two O and three F atoms, one of them stemming from the solvent molecule, in the corners. Two hfip F atoms are additionally coordinated to Li, albeit with longer distances; in summary, one hfip ligand is threefold coordinated, one twofold, and one once. The structure is repeated along a C_2 axis which intersects the C71-C71 and C73-C73 bonds.

There are three different kinds of Nb-O-C angles: 134.1±1.8°, 147.0±0.6°, and 165.3±0.9°. The latter value would hint to an almost ionic Nb-O bond. The smallest values belong to the ligands, which coordinate Li via their O atoms. Similarly, the Nb-O bond length for these two ligands is 206.0±1.2 pm, while for the other ligands, it is 191.1±2.4 pm.

One -C(CF₃)₂H group is disordered by rotation around the Nb-O axis. Hydrogen atoms were constructed.

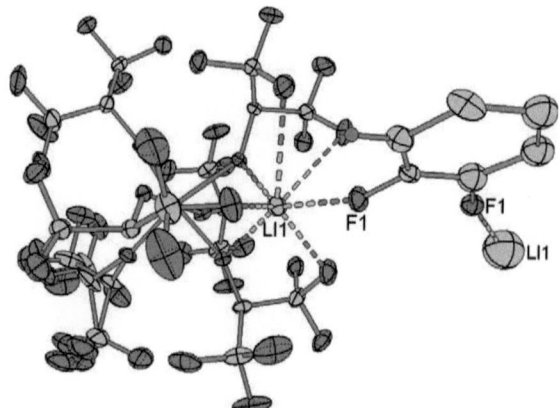

Figure 36: Asymmetric unit of [Li$_2$(1,2-C$_6$H$_4$F$_2$)][Nb(hfip)$_6$]$_2$ with completed solvent molecule and additional Li atom drawn to clarify the bridging. Thermal ellipsoids drawn at the 30% probability level. Hydrogen atoms omitted for clarity.

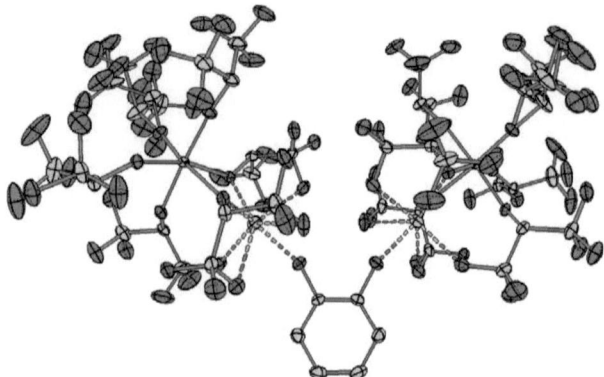

Figure 37: The two asymmetric units of [Li$_2$(1,2-C$_6$H$_4$F$_2$)][Nb(hfip)$_6$]$_2$ forming one superstructure bridged by the solvent molecule. Thermal ellipsoids drawn at the 30% probability level. Hydrogen atoms omitted for clarity.

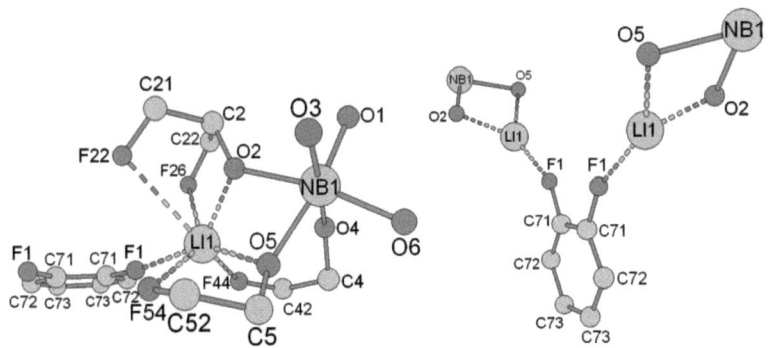

Figure 38: Section of the crystal structure of [Li₂(1,2-C₆H₄F₂)][Nb(hfip)₆]₂. The left side shows the distances d(Li-O) = 197.8±5.5 pm, d(Li-F) = 280.3±7.6 pm for the long contacts (to F22 and F26); 213.9±9.1 pm for the short contacts (to all other F's). The right side shows the bridging of two anions via Li atoms and the solvent (1,2-difluorobenzene).

By cation metathesis, Li⁺ in the aforementioned compound was exchanged for Ag⁺. Crystallization took place by cooling a saturated solution of the product in C_6H_5F to -20 °C. In the resulting structure (Figure 40; space group: P 2₁/c, R₁ = 3.79%), there are two independent units. In both, Ag is coordinated to the three most electron-rich C atoms of C_6H_5F with very different lengths (d(Ag-C) = 235.4...316.4 pm, ⌀ 261.8 pm) and to three oxygen atoms of the anion (d(Ag-O) = 239.6...269.8 pm, ⌀ 250.4 pm); see also Figure 39. The two units are twisted against each other; the torsion angle Ag1-Nb1-Nb2-Ag2 is -110.3°, while the torsion angle F101-Ag1-Ag2-F102 is -23.4°.

The angles Ag2-C203-F201 and Ag1-C104-F101 are 90.8° and 90.3°, meaning that the coordination of the solvent molecules is is almost perpendicular.

The ligands with coordinating O atoms have a Nb-O distance of 200.7±2.1 pm, while for the other ligands, it's 189.9±0.7 pm. The Nb-O-C angles are 137.0±3.3° and 157.4±4.3°, again with the smaller angles at the coordinating O atoms.

The crystallographic independence of the two units is verified by relatively large differences of the bond lengths as well as the conformation of the solvent molecule. The structure is twinned, the twin matrix $\begin{pmatrix} 0 & 0 & 1 \\ 0 & -1 & 0 \\ 1 & 0 & 0 \end{pmatrix}$ is common for a monoclinic space group and is the equivalent of a rotation around the bisector of a and c. The second domain has a share of 46.69%. No disordering was observed. All hydrogen atoms were constructed.

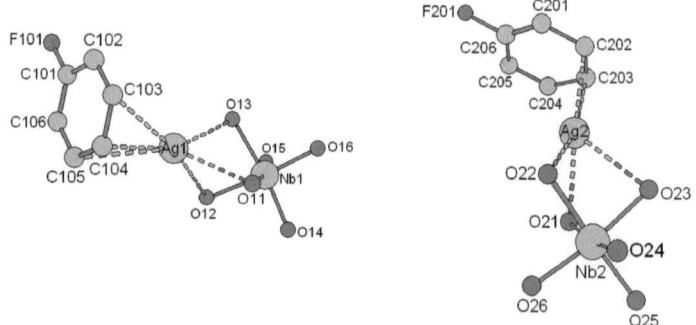

Figure 39: Section of the crystal structure of [Ag(C$_6$H$_5$F)][Nb(hfip)$_6$] showing the coordination of the Ag atoms.

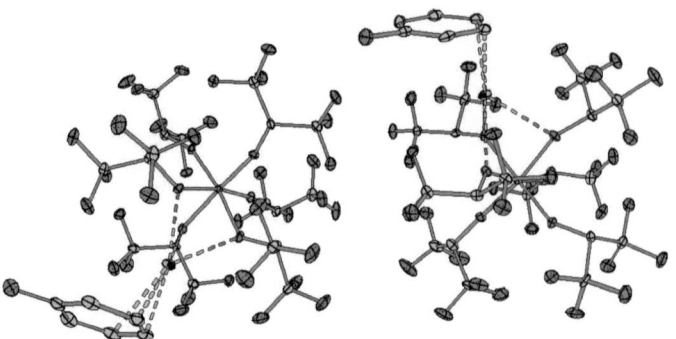

Figure 40: The asymmetric unit of [Ag(C$_6$H$_5$F)][Nb(hfip)$_6$]. Thermal ellipsoids drawn at the 30% probability level. Hydrogen atoms omitted for clarity.

The calculated (BP86/SV(P)) Nb-O-C angle of to [Nb(hfip)$_6$]$^-$ is 144°, the calculated Nb-O distance 200 pm (see Table 8). A similar ion can be found in the crystal structure of Be[Ta(OC(CH$_3$)$_2$H)$_6$]$_2$ (space group: P n).[46] Here, the Be atom is connected to two anions via two ligand O atoms each, giving a tetrahedral coordination with an average Be-O distance of 162.5±3.9 pm. The Ta-O distances can be grouped around 183.9±2.7, 195.0±3.2, and 212.2±3.3 pm, where the longest bonds (the third group) are formed by the O atoms which are coordinated to Be. The Ta-O-C angles can also be segmented into three groups: 138.3±4.6°, 155.7±2.5°, and 167.3±0.6°. The first group contains all angles with the Be-coordinating O atoms.

The attempt to synthesize [C$_4$MIm][Fe(hfip)$_4$] as a potential magnetic ionic liquid from [C$_4$MIm][FeCl$_4$] and four equivalents of Li-hfip led to a dark brown liquid from which a single brown crystal could be isolated. It proved to be [C$_4$MIm]$_2$[Fe$_2$(hfip)$_6$O], most likely formed by hydrolysis due to insufficient drying of the precursor IL. In this structure (Figure 43; space group: P $\bar{1}$, R$_1$ = 5.89%), two Fe(hfip)$_3$ moieties are linked by a single O atom, where the Fe-O-Fe angle is 177.3° (Figure 42). The Fe-O distance is 177.1±1.7 pm for the central O atom and 188.1±1.9 pm for the other O atoms, probably due to sterical crowding. There are two sets of Fe-O-C angles, grouped around 124.3±2.3° and 133.1±1.6°. Two -C(CF$_3$)$_2$H groups are disordered by variation of the Fe-O-C angle.

In the comparable structure bis(triethylammonium) (μ$_2$-oxo)-bis((2,2',2''-nitrilotribenzoato)-iron(III)) (space group: P $\bar{3}$c1), the Fe-O-Fe angle is 180°; the ligand Fe-O-C angles are 134.3° each. The Fe-O distance is 176.9 pm and 190.5 pm for the bridging and ligand O atoms, respectively.[47]

The cation assumes two positions, where one is translationally disordered, making a total of three conformers (Figure 43). The second cation is formed by rotation around 180° along the N-N axis of the first cation and translation perpendicular to the ring plane of roughly 350 pm.

All hydrogen atoms were added. Disordering included, there are five close H-O contacts (shortest: 209.5 pm) and 13 close H-F contacts (shortest: 170.2 pm).

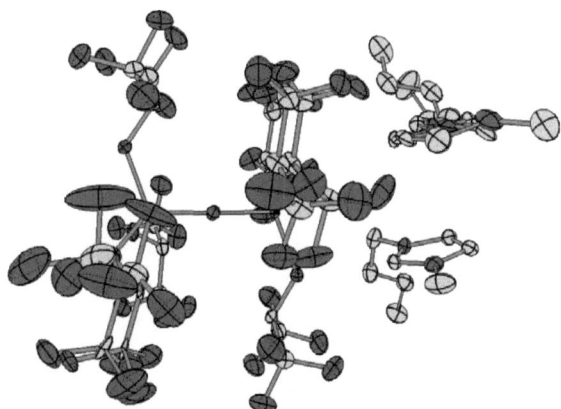

Figure 41: Asymmetric unit of [C$_4$MIm]$_2$[Fe$_2$(hfip)$_6$O]. Thermal ellipsoids drawn at the 30% probability level. Hydrogen atoms omitted for clarity.

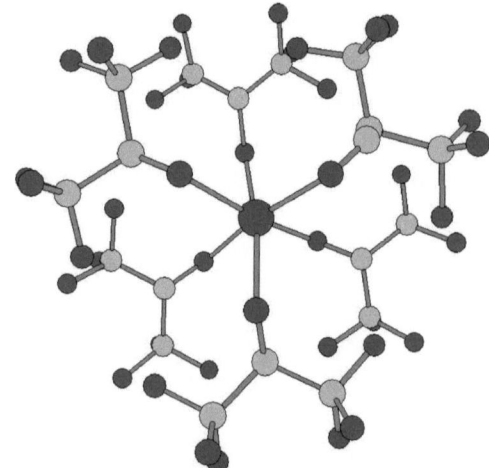

Figure 42: Ball-and-stick model of the anion in the lattice of [C$_4$MIm]$_2$[Fe$_2$(hfip)$_6$O] showing all six hfip ligands; the visible center is formed by one Fe atom, the bridging O and second Fe atom lie directly underneath. Hydrogen atoms omitted for clarity.

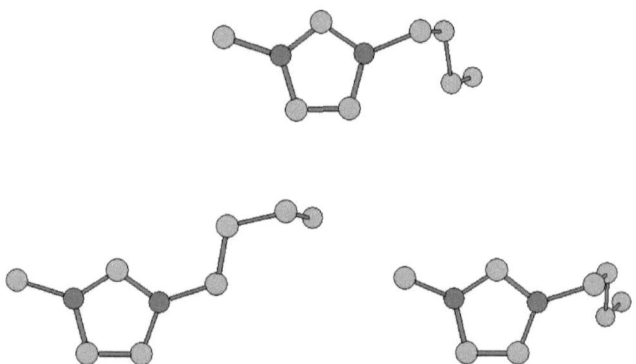

Figure 43: Ball-and-stick model of the different conformations of the cation in the lattice of [C$_4$MIm]$_2$[Fe$_2$(hfip)$_6$O]. Top: First cation; bottom: second cation in its two disordered forms. Hydrogen atoms omitted for clarity.

Structures of crystalline ionic liquids

Four ILs were crystallized from commercially available materials, yielding previously unknown structures. All hydrogen atoms in all four structures were added except for [C$_2$MIm][OTos] where all but H6A, H6B and H6C were refined. All structures are free of solvent molecules.

In the unit cell of [C$_2$MIm][OTos] (Figure 44; space group: P 2$_1$/c, R$_1$ = 3.89%), there is a total of 18 close H-O contacts per unit cell (i.e. 2.3 per ion pair, since Z = 8), forming the three-dimensional network shown in Figure 45; the shortest H-O contact is 227.5 pm. The crystal structure of [C$_4$MIm][OTos] (space group: P $\overline{1}$) is quite similar; there's a total of 3 H-O contacts per ion pair while the shortest H-O distance is 224.2 pm.[48]

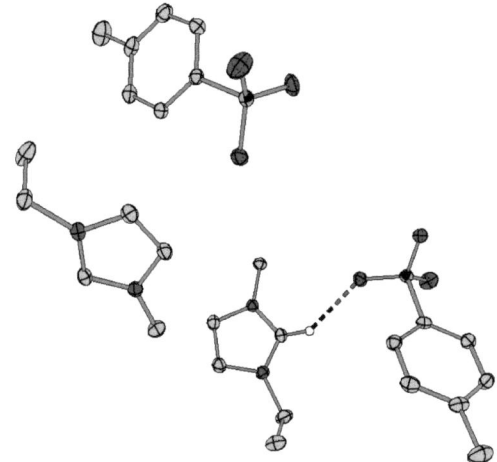

Figure 44: Asymmetric unit of [C$_2$MIm][OTos]. Thermal ellipsoids drawn at the 30% probability level. Non-coordinating hydrogen atoms omitted for clarity.

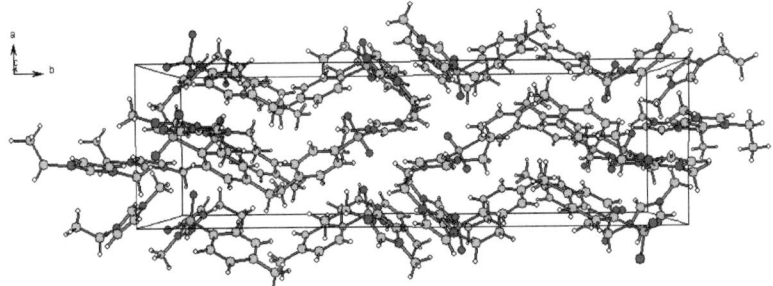

Figure 45: Packing diagram of [C$_2$MIm][OTos].

[N$_{4,4,4,4}$][BF$_4$] (Figure 46; R$_1$ = 6.36%) crystallizes in the rare space group P 2/c with 6 formula units in the asymmetric unit, where one [BF$_4$]$^-$ ion assumes two special positions (possibly of the Wyckhoff kind). Because of the cation assuming six different conformations (shown in Figure 48), the solution appears to be correct. This statement is supported by XPREP's proposal of P 2/c as well as PLATON's routine NEWSYM giving a probability of 62:38 of P 2/c against P c. However, NEWSYM suggests the existence of twinning or of a pseudo-symmetry, neither of which could be found during the refinement, though.

A multitude of conformations of a molecule in the same crystal lattice is relatively rare. A comparable structure is [N$_{3,3,3,3}$][O$_3$] (space group: P 2$_1$/c), which also holds six different conformers of the cation in its unit cell.[49] In the unit cell of [N$_{4,4,4,4}$][BF$_4$], there are 110 close H-F contacts – i.e. 4.6 per ion pair – being as short as 232.5 pm, forming a three-dimensional network (see Figure 47).

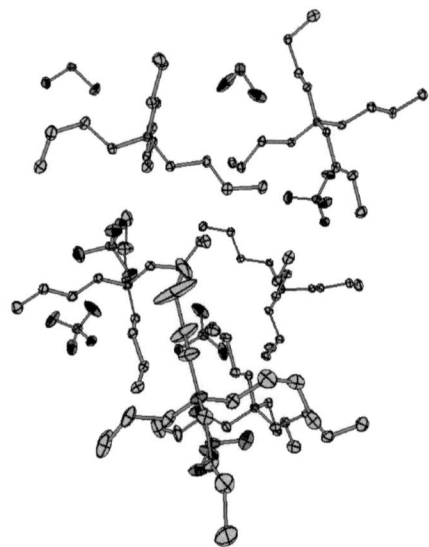

Figure 46: Asymmetric unit of [N$_{4,4,4,4}$][BF$_4$]. Thermal ellipsoids drawn at the 30% probability level. Hydrogen atoms omitted for clarity.

47

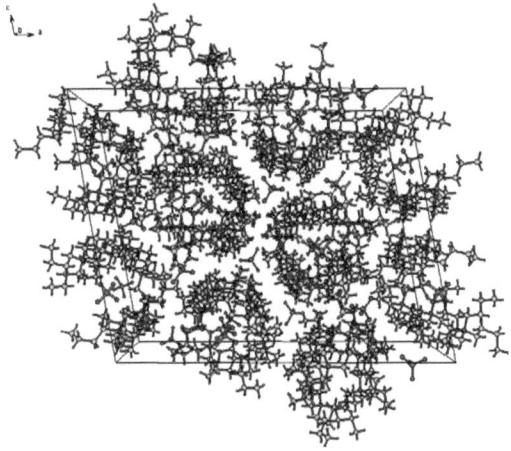

Figure 47: Packing diagram of [N$_{4,4,4,4}$][BF$_4$].

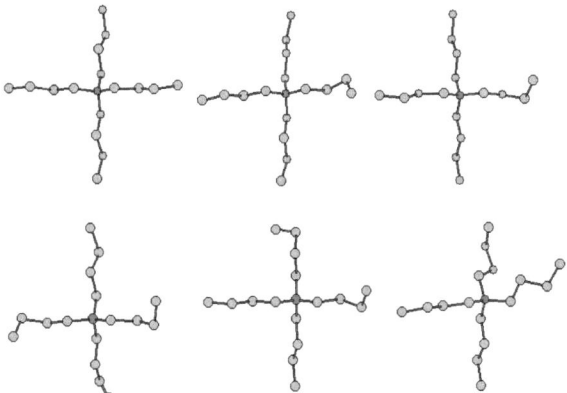

Figure 48: Ball-and-stick model of the six different conformations of the cation in the lattice of [N$_{4,4,4,4}$][BF$_4$]. Hydrogen atoms omitted for clarity.

The presence of different conformations of the cation in both [C$_4$MIm]$_2$[Fe$_2$(hfip)$_6$O] and [N$_{4,4,4,4}$][BF$_4$] led us to attempt a comparison to the calculated (BP86/TZVP) conformer space in the gas phase (cf. section 2.1.4). However, in the case of [C$_4$MIm]$^+$, no rmse of experimental and calculated structures < 0.2 nm could be found, whilst a good agreement would have an rmse ≤ 0.05 nm (using the program theseus as mentioned in section 2.1.4). The lowest-energy structure depicted in Figure 4 shows the best agreement still with all conformers in the lattice. For [N$_{4,4,4,4}$]$^+$, 98 different conformers were optimized in the gas phase; however, none gave an rmse to any of the conformations shown in Figure 48 better than 0.369 nm. The same thing holds true for COSMO geometries.

The explanation is that conformers in the crystal lattice don't have to assume minima as they do in the gas phase. This goes for both atomic distances and angles; atoms lie on different positions in the lattice such that small deviations can sum up, making any agreement with gas phase structures impossible.

The structures of $[N_{1,1,1,1}][OTfa]$ and $[S_{1,\varphi,\varphi}][PF_6]$ (Figure 49 to Figure 52; space groups: P 2_1/m and P $\bar{1}$, R_1 = 3.05% and 3.57%, respectively) are the ones of common organic salts. $[N_{1,1,1,1}][OTfa]$ could be compared to $[N_{1,1,1,1}][OAc]$ (space group: P $\bar{1}$);[50] however, the latter contains four water molecules per ion pair, the former none. In $[N_{1,1,1,1}][OTfa]$, two of the three fluorine atoms are disordered; in $[S_{1,\varphi,\varphi}][PF_6]$, all of them. Its structure is similar to $[S_{1,\varphi,\varphi}][SbF_6]$ (same space group);[51] the torsion angle C4-S1-C7-C13 is only slightly smaller (107.6° vs. 108.4°). However, the phenyl rings are twisted in a different way, since the torsion angle C5-C4-S1-C7 is -40.5° vs. 64.0°.

In the case of $[S_{1,\varphi,\varphi}][PF_6]$, there are five close H-F contacts per ion pair, being as short as 231.8 pm. In contrast, the ion pair found in $[N_{1,1,1,1}][OTfa]$ forms four long H-O contact with at least 247.2 pm length.

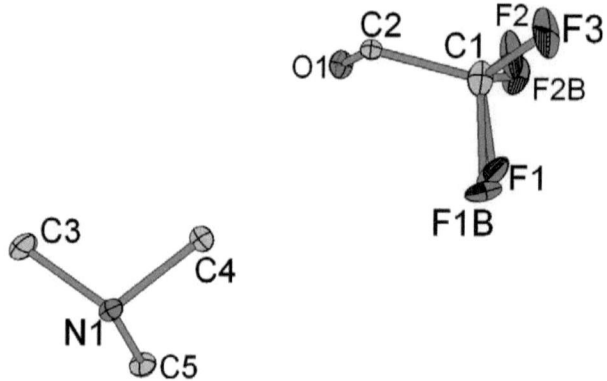

Figure 49: Asymmetric unit of $[N_{1,1,1,1}][OTfa]$. Thermal ellipsoids drawn at the 30% probability level. Hydrogen atoms omitted for clarity.

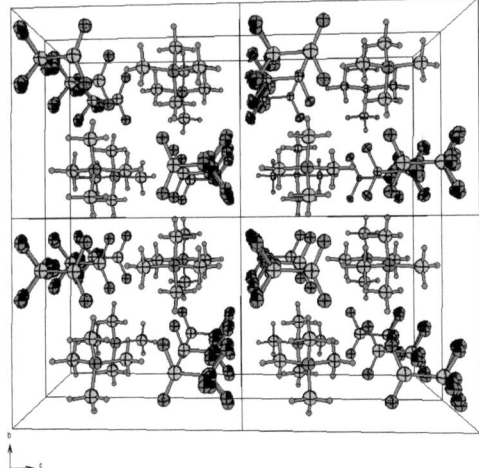

Figure 50: Packing diagram of [N$_{1,1,1,1}$][OTfa].

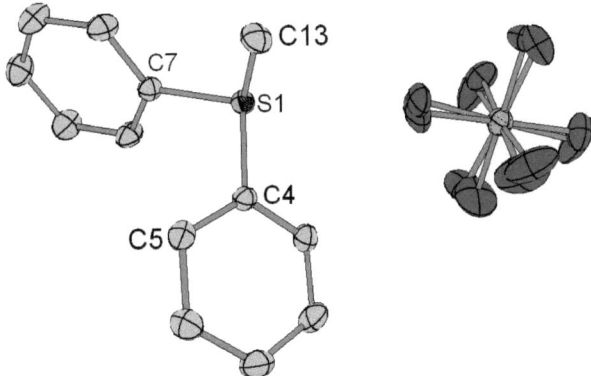

Figure 51: Asymmetric unit of [S$_{1,\varphi,\varphi}$][PF$_6$]. Thermal ellipsoids drawn at the 30% probability level. Hydrogen atoms omitted for clarity.

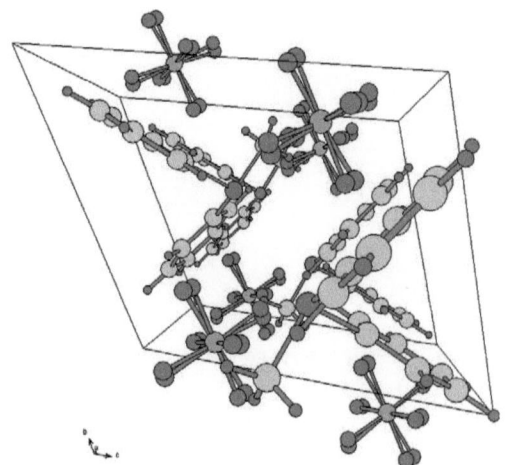

Figure 52: Packing diagram of [S$_{1,\varphi,\varphi}$][PF$_6$].

3. Weakly Coordinating Anions

3.1. General Introduction

Anions in which the negative charge is delocalized over a comparatively large surface don't form strong ionic bonds to their counterions, but weak dispersive interactions instead.[52,53] Therefore, they are called "weakly coordinating". Since the lattice energy of an ionic compound scales reciprocal to its molecular volume, compounds that contain these large anions tend to become soluble in solvents of low polarity.

Because of their low basicity, they may stabilize delicate cations that were previously known only to be stable in the gas phase. Thus, it is generally desired that these anions are stable against electrophilic attack and oxidation as well as reduction. To find a compound that fits all these criteria and, at the same time, is non-toxic, cheap and easy to synthesize is next to impossible. However, we suggest identifying "task-bound" WCAs, i.e. a limited number of compounds that can ideally be used for one specific purpose. Certain crucial parameters of these compounds are accessible via quantum chemistry, as we will show, so a screening can take place without tedious attempts at synthesis and characterization.

3.2. Computational Study of $[M(CF_3)_4]^-$-type WCAs[54]

3.2.1. Background

In 1986, Naumann and Dukat prepared the first tetrakis(trifluoromethyl)-coinage metallate(III), Ag[Ag(CF$_3$)$_4$], by a disproportionation reaction of Ag(I) compounds.[55] The lighter homologue [Cat][Cu(CF$_3$)$_4$] (Cat$^+$ = bulky cation like [PNP]$^+$ = [Ph$_3$PNPPh$_3$]$^+$) was first prepared in 1993 in a simple procedure by oxidation of CuCF$_3$.[56] The heavier and more difficult to obtain [Au(CF$_3$)$_4$]$^-$ compounds were also synthesized in 1993.[57] All tetrakis(trifluoromethyl)-coinage metallates(III) were found to be stable against air, moisture or elevated temperatures (decomposition point: 140 °C for [PNP][Cu(CF$_3$)$_4$]) and are readily soluble in organic solvents. However, their sensitivity to light increases in going from Cu to Au. The geometry of the anions is almost planar and the – at least for Cu and Ag – unusual oxidation state +III is stabilized by two layers of fluorine atoms that form above and below the MC$_4$ plane, shielding the metal atom against nucleophilic attack. These coinage metallates were used as weakly coordinating anions to prepare cationic organic superconductors with a transition temperature of up to 11.1 K at ambient pressure.[58] This ability was attributed to the flat, highly symmetrical geometry of the WCAs, allowing close crystal packing in sheets without disordering; and their low charge-to-volume ratio, being beneficent to Cooper pair formation.

In this contribution, we investigated the properties of the [M(CF$_3$)$_4$]$^-$ WCAs with earlier introduced computational models,[59] and the quality of these anions in comparison to other frequently used WCAs is established.

3.2.2. Results

In order to test the reliability of the computational models for the investigation of the title compounds, we compared the experimental and calculated structures (see Table 23) and found them to be in excellent agreement, even at the relatively inexpensive level of theory used (BP86/SV(P)). Inclusion of solvation by the COSMO model, approximating the polar solvent 1,2-difluorobenzene (ε_r = 13.38), led to only minor changes in the structures.

Table 23: Comparison of experimental[60] and computed structural data of [M(CF$_3$)$_4$]$^-$ WCAs. In all experiments, the cation was [PNP]$^+$.

	[Cu(CF$_3$)$_4$]$^-$		[Ag(CF$_3$)$_4$]$^-$		[Au(CF$_3$)$_4$]$^-$	
	exp.	calc.	exp.	calc.	exp.	calc.
symmetry	D_{2d}	D_{2d}	D_{2d}	D_{2d}	D_{2d} [a]	D_{2d}
d(M-C) [pm]	197	202	208	214	211±5	214
d(C-F) [pm]	135±3	137	134	137	128±8	137
∠ C-M-C, small [°]	91	91	90	90	90	90
∠ C-M-C, large [°]	169	168	173	172	175	173

[a] Distorted.

The calculated structures of the species [M(CF$_3$)$_4$]$^-$ are in a flat minimum on the potential energy surface, as the CF$_3$ units are easily rotated around the M-C bond. The anions therefore exist in D_{2d} or C_{4h} symmetry; both are local minima with a ΔH^{0K} of 2.5 kJ mol^{-1} at most.

A complete assignment of the calculated bands of the title compounds is given in Table 24 to facilitate further synthetic uses. The comparison of calculated and experimental vibrational spectra shows generally very good agreement.

Since theory describes the structures of the coinage metallates [M(CF$_3$)$_4$]$^-$ rather well, we are confident that the calculated geometries of the, as to now, poorly studied or completely unknown compounds M(CF$_3$)$_3$, [M(CF$_3$)$_3$F]$^-$, and M(CF$_3$)$_3$(CF$_2$) will also be close to the experimentally expected values. These molecules are needed to study the WCA properties of the coinage metallates, so we briefly present them in Figure 53.

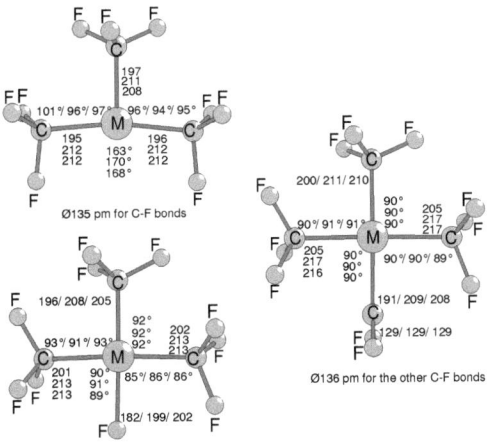

Figure 53: Calculated gas phase structures of M(CF$_3$)$_3$, M(CF$_3$)$_3$(CF$_2$) and [M(CF$_3$)$_3$F]$^-$ and their principal structural parameters at the BP86/SV(P) level. The first value addresses the Cu, the second the Ag and the third the Au compounds. Bond lengths are given in pm. All species represent global minima in C_1 symmetry.

Which WCA should be used to synthesize and investigate the cation in question is determined by the properties of the latter. There is at present no WCA that fits every purpose. However, we recently proposed a quantum chemical set of criteria to base a decision for a particular WCA on hard numbers (Figure 54):[59,61]

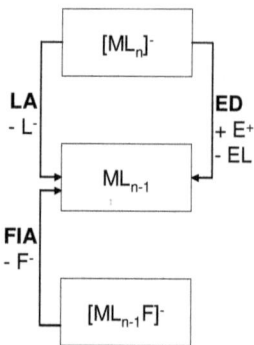

Figure 54: Reactions used to computationally assess the quality of [ML$_n$]$^-$ type WCAs.

In the following we investigate the properties of the coinage metallates as WCAs according to the subsequent rationales:

- ED (electrophile-induced decomposition): The energy of the reaction of a WCA with an electrophile like H$^+$, Cu$^+$, or [SiMe$_3$]$^+$, abstracting a ligand anion. The reaction energies are specifically denoted PD, CuD, and TMSD. The ED family of values is indicative of the WCA's resilience against hydrolysis as well as hard and soft Lewis acids. We assessed the TMSD with and without inclusion of COSMO solvation energies approximating a polar 1,2-difluorobenzene solution.

- FIA (fluoride ion affinity) and LA (ligand affinity): The energy of the reaction of the WCA's corresponding Lewis acid with a fluoride and a ligand anion, respectively. The FIA is indicative for the strength of the Lewis acid. The LA is indicative of the WCA's resilience against ligand abstraction, yielding a free ligand ion.

- HOMO and HOMO-LUMO gap: The lower the HOMO value, the harder the WCA is to oxidize; the larger the HOMO-LUMO gap, the harder the WCA is to reduce.

- q_{neg} and q_{surf}: the partial charge on the most negative atom of the entire WCA as well as its accessible surface, respectively. The closer to zero they are, the lower the WCA's coordination tendency.

Table 24: Calculated vibrational frequencies and assignments of the bands of [M(CF$_3$)$_4$]$^-$; comparison to Ag[Ag(CF$_3$)$_4$]. Intensities are given in km mol^{-1}.

	Cu				Ag				Au				Ag, exp.[62]
sym.	mode [a]	int. IR [cm^{-1}]	Raman active?	sym.	mode [a]	[cm^{-1}]	int. IR	Raman active?	sym.	mode [a]	[cm^{-1}]	int. IR	Raman active?	
a$_2$	τ(CF$_3$)	21	0	no	a$_2$	τ(CF$_3$)	23	0	no	a$_2$	τ(CF$_3$)	18	0	no	
a$_1$	γ$_s$(MC$_4$)	42	0	yes	a$_1$	γ$_s$(MC$_4$)	45	0	yes	e	τ(CF$_3$)	48	0	yes	
e	τ(CF$_3$)	53	0	yes	e	τ(CF$_3$)	48	0	yes	a$_1$	γ$_s$(MC$_4$)	50	0	yes	
b$_1$	τ(CF$_3$)	66	0	yes	b$_1$	τ(CF$_3$)	56	0	yes	b$_1$	τ(CF$_3$)	56	0	yes	
b$_2$	γ$_s$(MC$_4$)	92	1	yes	b$_2$	γ$_s$(MC$_4$)	84	1	yes	b$_2$	γ$_s$(MC$_4$)	84	1	yes	
e	δ$_{as}$(MC$_4$)	122	1	yes	e	δ$_{as}$(MC$_4$)	111	0	yes	e	δ$_{as}$(MC$_4$)	115	0	yes	
b$_1$	δ$_s$(MC$_4$)	141	0	yes	b$_1$	δ$_s$(MC$_4$)	124	0	yes	b$_1$	δ$_s$(MC$_4$)	131	0	yes	
b$_2$	ν$_s$(MC$_4$)	187	0	yes	b$_2$	ν$_s$(MC$_4$)	190	0	yes	e	γ$_{as}$(MC$_4$)	210	1	yes	205 m, p
e	γ$_{as}$(MC$_4$)	188	0	yes	e	γ$_{as}$(MC$_4$)	191	0	yes	b$_2$	ν$_s$(MC$_4$)	212	0	yes	210 sh
a$_1$	ν$_s$(MC$_4$)	200	0	yes	a$_1$	ν$_s$(MC$_4$)	204	0	yes	a$_2$	γ$_{as}$(CF$_3$)	226	0	no	
e	γ$_{as}$(CF$_3$)	215	13	yes	e	γ$_{as}$(CF$_3$)	212	0	yes	a$_1$	ν$_s$(MC$_4$)	227	0	yes	
a$_2$	γ$_{as}$(CF$_3$)	218	0	no	a$_2$	γ$_{as}$(CF$_3$)	217	0	no	e	γ$_{as}$(CF$_3$)	228	26	yes	225 vs
b$_1$	γ$_{as}$(CF$_3$)	229	0	yes	b$_1$	γ$_{as}$(CF$_3$)	223	0	yes	b$_1$	γ$_{as}$(CF$_3$)	244	0	yes	
a$_1$	γ$_s$(MC$_4$)	246	0	yes	a$_1$	γ$_s$(MC$_4$)	244	0	yes	a$_1$	γ$_s$(MC$_4$)	268	0	yes	
b$_2$	γ$_s$(MC$_4$)	285	5	yes	b$_2$	γ$_s$(MC$_4$)	270	2	yes	b$_2$	γ$_s$(MC$_4$)	286	2	yes	
e	δ$_{as}$(MC$_4$)	342	14	yes	e	δ$_{as}$(MC$_4$)	300	18	yes	e	δ$_{as}$(MC$_4$)	287	8	yes	
a$_2$	δ$_{as}$(CF$_3$)	499	0	no	a$_2$	δ$_{as}$(CF$_3$)	496	0	no	a$_2$	δ$_{as}$(CF$_3$)	497	0	no	
b$_1$	δ$_{as}$(CF$_3$)	501	0	yes	b$_1$	δ$_{as}$(CF$_3$)	498	0	yes	b$_1$	δ$_{as}$(CF$_3$)	501	0	yes	
e	δ$_{as}$(CF$_3$)	503	0	yes	e	δ$_{as}$(CF$_3$)	502	0	yes	e	δ$_{as}$(CF$_3$)	507	0	yes	
e	δ$_{as}$(CF$_3$)	509	0	yes	e	δ$_{as}$(CF$_3$)	505	0	yes	e	δ$_{as}$(CF$_3$)	508	0	yes	
b$_2$	δ$_{as}$(CF$_3$)	513	0	yes	b$_2$	δ$_{as}$(CF$_3$)	511	0	yes	b$_2$	δ$_{as}$(CF$_3$)	516	0	yes	
a$_1$	δ$_{as}$(CF$_3$)	519	0	yes	a$_1$	δ$_{as}$(CF$_3$)	515	0	yes	a$_1$	δ$_{as}$(CF$_3$)	521	0	yes	
b$_2$	ν$_s$(MC$_4$)	687	2	yes	b$_2$	ν$_s$(MC$_4$)	687	1	yes	b$_2$	ν$_s$(MC$_4$)	692	1	yes	
a$_1$	ν$_s$(MC$_4$)	692	0	yes	a$_1$	ν$_s$(MC$_4$)	690	0	yes	e	ν$_{as}$(MC$_4$)	693	5	yes	720 s
e	ν$_{as}$(MC$_4$)	694	13	yes	e	ν$_{as}$(MC$_4$)	691	8	yes	a$_1$	ν$_s$(MC$_4$)	696	0	yes	
a$_2$	ν$_{as}$(CF$_3$)	991	0	no	a$_2$	ν$_{as}$(CF$_3$)	999	0	no	a$_2$	ν$_{as}$(CF$_3$)	998	0	no	
b$_2$	ν$_s$(CF$_3$)	1010	88	yes	b$_2$	ν$_s$(CF$_3$)	1026	46	yes	e	ν$_{as}$(CF$_3$)	1047	13	yes	
e	ν$_{as}$(CF$_3$)	1038	125	yes	e	ν$_{as}$(CF$_3$)	1043	45	yes	b$_2$	ν$_s$(CF$_3$)	1048	48	yes	1045 vs
e	ν$_{as}$(CF$_3$)	1048	252	yes	e	ν$_{as}$(CF$_3$)	1048	0	yes	e	ν$_s$(CF$_3$)	1053	0	yes	
a$_1$	ν$_{as}$(CF$_3$)	1051	0	yes	a$_1$	ν$_{as}$(CF$_3$)	1051	18	yes	a$_1$	ν$_{as}$(CF$_3$)	1063	130	yes	
e	ν$_{as}$(CF$_3$)	1075	1024	yes	e	ν$_{as}$(CF$_3$)	1076	1274	yes	b$_1$	ν$_{as}$(CF$_3$)	1083	0	yes	
b$_1$	ν$_{as}$(CF$_3$)	1079	0	yes	b$_1$	ν$_{as}$(CF$_3$)	1076	0	yes	e	ν$_{as}$(CF$_3$)	1093	1059	yes	
b$_2$	ν$_{as}$(CF$_3$)	1090	754	yes	b$_2$	ν$_{as}$(CF$_3$)	1081	846	yes	b$_2$	ν$_{as}$(CF$_3$)	1093	886	yes	
a$_1$	ν$_s$(CF$_3$)	1159	0	yes	a$_1$	ν$_s$(CF$_3$)	1145	0	yes	a$_1$	ν$_s$(CF$_3$)	1154	0	yes	1160 m

[a] s = symmetric, as = asymmetric.

In Table 25, the performance of the coinage metallates is compared to other frequently used WCAs that were assessed using the same methodologies. The FIA and ED values of the coinage metallates are competitive to those of the more common $[BF_4]^-$ and $[PF_6]^-$, which suggests a similar thermodynamic stability towards electrophilic attack. Especially the LA values are rather high and indicative for high stability. Similar HOMO values suggest similar stability against oxidation. In general, the performance of the coinage metallates tends to improve with increasing weight of the central atom, i.e. from Cu to Au.

For the tetrakis(trifluoromethyl)metallates(III), there are additional pathways of decomposition possible rather than ligand abstraction. This was explicitly shown for the related tetrahedral $K[B(CF_3)_4]$,[63] in which a F^- ion is abstracted by AsF_5.[64] The resulting $B(CF_3)_3(CF_2)$ was never isolated as it immediately underwent rearrangement. Also, $B(CF_3)_3$ was never made as it is much more Lewis acidic than BF_3 (see the FIA values in Table 25) and thus favors rearrangement. $[Ag(CF_3)_4]^-$, on the other hand, was shown to react with Lewis acids in CH_3CN to give $Ag(CF_3)_3(CH_3CN)$, which in turn exchanges CH_3CN for other ligands.[65] The argentate was also observed to release CF_2 (and likely HF) with strong Brønsted acids. Such decomposition pathways have to be accounted for by the computational models.

In $[B(CF_3)_4]^-$, the compact tetrahedral geometry shields the boron atom from electrophilic attack. The planar geometry of the coinage metallates seems to be a disadvantage in this respect (see Figure 55), but the electrostatic repulsion from the positively polarized metal (and carbon) atoms, as expected from first principles (e.g. electronegativity), would render electrophilic attack from above the MC_4 plane unlikely.

Figure 55: Calculated space-filling model of $[Ag(CF_3)_4]^-$, showing the planar geometry.

In agreement with the experimental data available, we thus propose the following for $[M(CF_3)_4]^-$: An attacking electrophile E^+ may likely not break the M-C bond, but would initially abstract a F⁻ ion from one of the CF₃ groups. In the case of M = B, the resulting carbene complex is more tightly bound (see below, Table 26) and remains coordinated until it inserts into a neighboring C-F bond; in the case of M = Cu, Ag, Au, the CF_2 moiety may eventually dissociate from $M(CF_3)_3$ and react with E-F to give E-CF₃.

Therefore, we studied several novel modes of decomposition involving a CF_2 carbene complex (Figure 56).

The computational access to the investigated quantities uses isodesmic reactions for the larger systems and experimental or reliable MP2/TZVPP ab initio values for the smaller systems. Due to error cancellation we expect these values to be accurate within 20 kJ mol⁻¹; relative trends are definitely correct as only one methodology was used for all WCAs. The following rationale is behind the equations sketched in Figure 56:

FIAC ("fluoride ion affinity to carbene"): The energy of the reaction of $[M(CF_3)_4]^-$ yielding the carbene complex $M(CF_3)_3(CF_2)$ and a free F⁻ ion.

CED ("carbene-retaining electrophile-induced decomposition"): Derived from the FIAC are the energies of the reaction of $[M(CF_3)_4]^-$ with an electrophile like H⁺, Cu⁺, or [SiMe₃]⁺, yielding the fluorides of these electrophiles[66] as well as $M(CF_3)_3(CF_2)$. They are specifically abbreviated as CPD, CCuD, and CTMSD.

CA ("carbene affinity"): The energy of the reaction of $M(CF_3)_3(CF_2)$ yielding a free CF_2 molecule and $M(CF_3)_3$.

EFD ("electrophile fluoride decomposition"): Derived from the CA are the energies of the reaction of $M(CF_3)_3(CF_2)$ with the electrophile fluorides HF, CuF,[66] and Me₃SiF, yielding E-CF₃ as well as $M(CF_3)_3$. The reaction energies are specifically abbreviated as HFD, CuFD, and TMSFD.

All entries are included in Table 26 below.

Table 25: [M(CF$_3$)$_4$]$^-$ in comparison to other known WCAs.

WCA	sym.	FIA [kJ mol^{-1}]	LA [kJ mol^{-1}]	PD [kJ mol^{-1}]	CuD [kJ mol^{-1}]	TMSD [a] [kJ mol^{-1}]	HOMO [eV]	HOMO-LUMO gap [eV]	q$_{neg}$	q$_{surf}$
[Cu(CF$_3$)$_4$]$^-$	D_{2d}	374	381	-1273	-512	-483/ -138	-2.362	3.099	-0.90 Cu [b]	-0.20 F
[Ag(CF$_3$)$_4$]$^-$	D_{2d}	367	390	-1263	-503	-474/ -136	-2.381	3.565	-0.21 F	-0.21 F
[Au(CF$_3$)$_4$]$^-$	D_{2d}	383	420	-1233	-473	-443/ -108	-2.547	4.667	-0.21 F	-0.21 F
[B(CF$_3$)$_4$]$^-$ [c]	T	554	520	-1133	-373	-343/ 2	-3.532	9.078	-2.14 B	-0.10 F
[BF$_4$]$^-$ [59]	T_d	308	338	-1212	-521	-622/ -210	-1.799	10.820	-0.25 F	-0.25 F
[PF$_6$]$^-$ [59]	O_h	394	394	-1156	-465	-566/ -168	-2.672	8.802	-0.44 F	-0.44 F
[B(C$_6$F$_5$)$_4$]$^-$ [59]	S_4	444	296	-1256	-538	-488/ -187	-3.130	4.196	-0.21 F	-0.21 F
[Al(hfip)$_4$]$^-$ [59]	S_4	537	342	-1081	-395	-400/ -96	-4.100	6.747	-0.24 O	-0.20 F
[Sb(OTeF$_5$)$_6$]$^-$ [59]	C_3	633	341	-973	-353	-337/ -52	-6.610	2.326	-0.61 O	-0.39 F

[a] Gas phase/ 1,2-difluorobenzene solution; note that the reaction energy trends are about the same but are slightly favoring the coinage metallates in solution over the larger WCAs. [b] See computational details for explanation. [c] Ref. [58] gives slightly different values, except for LA, which is given as 490 kJ mol^{-1}.

Table 26: Calculated energies of the alternative decomposition reactions of [M(CF$_3$)$_4$]$^-$ as shown in Figure 56; values are in kJ mol^{-1}.

M	FIAC	CPD	CCuD	CTMSD [a]	CA	HFD	CuFD	TMSFD [b]
Cu	450	-1104	-378	-511/ -173	132	-129	-79	35
Ag	475	-1078	-352	-485/ -153	116	-145	-95	19
Au	485	-1069	-343	-476/ -151	137	-124	-74	40
B	511	-1043	-317	-450/ -111	211	-50	0	114

[a] Calculated in gas phase/ 1,2-difluorobenzene solution. As in Table 25, inclusion of the solvation model gives almost identical trends as the gas phase values, indicating that the latter can be used for the assessment of the stability of a given WCA. [b] The TMSFD values were not calculated in solution, since only uncharged particles are involved in the reaction.

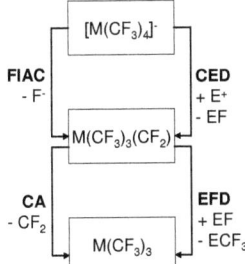

Figure 56: Overview of the additional decomposition pathways of tetrakis(trifluoromethyl)-metallates(III).

In case of E$^+$ = H$^+$ and Cu$^+$, the CED energies imply strongly exothermic reactions, albeit not as exothermic as the ED reactions, meaning that an attacking electrophile E$^+$ would rather abstract a [CF$_3$]$^-$ than an F$^-$ ion. But, as explained above, this reaction may be impossible because of sterical or electrostatic reasons and therefore it appears that kinetics is in favor of a decomposition pathway as delineated in Figure 56. The intermediate formation of a carbene complex would also explain the release of free CF$_2$,[65] as a strong donor solvent like CH$_3$CN could exchange with it as a ligand. In the case of E$^+$ = [SiMe$_3$]$^+$, the abstraction of an F$^-$ ion is more exothermic than the abstraction of a [CF$_3$]$^-$ moiety (CTMSD < TMSD), probably because of the particularly high stability of the Si-F bond.

To explain the eventual abstraction of CF$_2$ from the carbene complexes with M = Cu, Ag, Au, but not B, we consider the HFD and CuFD values: It becomes evident that, for the coinage metallates, an initially created fluoride E-F could further exothermically react with a coordinated CF$_2$ moiety of the carbene complex. By contrast, for M = B, this reaction is much less favored so that the observed rearrangements can take place instead.

In case of $E^+ = [SiMe_3]^+$, further abstraction of the remaining CF_2 moiety would be endothermic and unlikely to occur (TMSFD > 0).

3.2.3. Conclusion

Table 25 shows that the title compounds are at least in terms of thermodynamic stability competitive to $[PF_6]^-$. It appears that kinetics further stabilize these WCAs to a very comfortable level. To illustrate this: The small HOMO-LUMO gap of the coinage metallates suggests susceptibility to reduction, meaning that these anions could act as oxidizers towards their reaction partners. Due to kinetic hindrance, though, the redox potential of the system $[Ag(CF_3)_4]^-/[Ag(CF_3)_2]^-$ in CH_3CN is -1.25 V, which shows that rather strong reducing agents would be needed to interfere with the argentate, even though it is an Ag(III) compound.[65]

Another aspect is the observed decomposition by an attacking reactive electrophile. Although, as Table 25 and Table 26 show, the CuD and PD values are considerably larger than those of CCuD and CPD, the electrophile would not abstract a CF_3 ligand due to kinetics. Thus, to decide on the quality of the coinage metallates to be used as a WCA for a specific problem, the kinetic values collected in Table 26 should rather be used. They are more competitive to other, first-class WCAs and suggest that the coinage metallates should be used more widely. This also follows from model calculations with inclusion of solvation (COSMO, 1,2- $C_6H_4F_2$) and reiterates our notion that the stability of a WCA may be analyzed based on calculated gas phase quantities.[59]

Furthermore, the flat, compact form of the tetrakis(trifluoromethyl)-coinage metallate(III) anions allow for highly ordered, dense crystal packing. The easy syntheses, the trouble-free handling of the compounds in air, the thermal stability and the low coordination tendency show high potential for further research.

3.2.4. Appendix: Additional Computational Details

To calculate energies, $M(CF_3)_3$, $[M(CF_3)_3F]^-$, $M(CF_3)_3(CF_2)$, and $[M(CF_3)_4]^-$ were optimized with the highest possible symmetry at the BP86/SV(P)[4,5,6,7] level at 0 K in gas phase and in 1,2-difluorobenzene solution ($\varepsilon_r = 13.38$; modeled with COSMO[13]), respectively. Vibrational analyses were carried out for every molecule to make sure they represent local minima. Isodesmic reactions were calculated at the BP86/SV(P) level, non-isodesmic reactions at the MP2/TZVPP level of theory.[67,12,68,69] All MP2 optimizations were performed using frozen core extrapolation for the inner shells. Quantities that were derived in analogy to the values originally published in ref. [59] were assessed similar to in the original publication. For the heavy elements Ag and Au, the core electrons were replaced by a quasi-relativistic electrostatic core potential with 28 and 60 electrons, respectively.[70] This treatment was shown to be reliable for a multitude of Ag compounds prepared and calculated in our group. DFT and MP2 calculations were carried out with the TURBOMOLE[8] program package using the Resolution of Identify (RI) approximation.[9]

Partial charges were calculated using the Population Analysis Based On Occupation Numbers (PABOON).[71] The partial charge on the metal atoms (Cu, Ag, Au) lies between -1 and +2, depending very strongly on the number of MAOs employed for the calculation. To stay compatible with our previous study, we used the standard selection threshold and the default number of MAOs chosen by the routine MOLOCH.[59] It is, however, important to note that in contrast to the charges of metal atoms, the partial charges on the fluorine atoms always stays between -0.15 and -0.25, no matter the number of MAOs used for the analysis. Thus, we consider the partial charges on the metal atoms in $[M(CF_3)_4]^-$ as not very reliable; first principle approaches like electronegativity suggest them to be positive (cf. $\chi_{AR}(Cu) = 1.8$, $\chi_{AR}(Ag) = 1.4$, $\chi_{AR}(Au) = 1.4$ and the group electronegativity of CF_3 being around 3.3 to 3.5).

3.3. Computational Screening of Further Possible Weakly-Coordinating Anions

3.3.1. Background

Based on the results above, we performed a screening on possible WCA candidates to decide for which a synthesis would be justified. We used the ligands given in Figure 57 which are – in some basic form – either commercially directly available or easy to prepare (see Table 27).

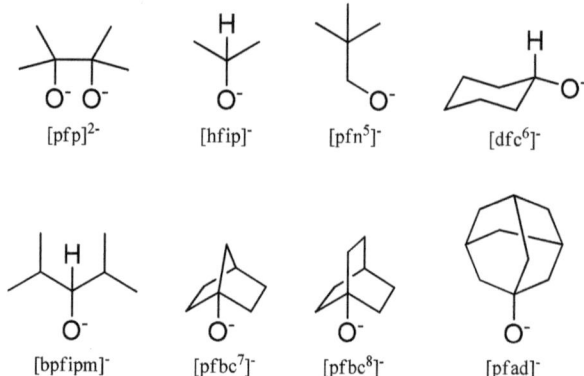

Figure 57: Legend for fluorinated ligands, sorted with increasing sterical hindrance. Note that all free sites at the carbon atoms are occupied with F, not H. Substituting the proton of [dfc⁶]⁻ with CH₃ leads to [mpfc⁶]⁻. Substituting the proton of [hfip]⁻ with CH₃ leads to [hftb]⁻, with CF₃ to [pftb]⁻.

Table 27: Ligands, their starting materials and prices (as given on 20/04/09 from P&M-Invest, Moscow, Russia, except for perfluoroadamantanol which is available on request from Idemitsu Chemicals, Tokyo, Japan).

ligand	full name	formula	educt	price [€ mmol⁻¹]	safety
[pfp]²⁻	perfluoropinacodiyl	$C_6O_2F_{12}^{2-}$	alcohol[a]	0.68	Xi [c]
[hfip]⁻	bis(trifluoromethyl)methoxyl	$C_3HOF_6^-$	alcohol[a]	0.029	-/-
[pfn⁵]⁻	perfluoroneopentoxyl	$C_5OF_{11}^-$	alcohol		
[dfc⁶]⁻	1H-perfluorocyclohexoxyl	$C_6HOF_{10}^-$	keton[a]	2.7	-/-
[bpfipm]⁻	bis(perfluoroisopropyl)methoxyl	$C_7HOF_{14}^-$	keton[a]	2.9	Xi
[pfbc⁷]⁻	perfluorobicyclo[2.2.1]heptoxyl	$C_7OF_{11}^-$	alcohol		
[pfbc⁸]⁻	perfluorobicyclo[2.2.2]octoxyl	$C_8OF_{13}^-$	alcohol		
[pfad]⁻	perfluoroadamantoxyl	$C_{10}OF_{15}^-$	alcohol[a]	[b]	-/-

[mpfc⁶]⁻	1-methyl-1H-perfluorocyclohexyl	$C_7H_3OF_{10}^-$	same as [dfc⁶]⁻		
[hftb]⁻	bis(trifluoromethyl)-methyl-methoxyl	$C_4H_3OF_6^-$	alcohol[a]	0.054	Xi, F
[pftb]⁻	perfluorotert.butoxyl	$C_4OF_9^-$	alcohol[a]	0.25	Xn

[a] Commercially available starting material. [b] No price given. [c] Several biological studies suggest very high toxicity.

In the following nine tables, the calculated values for 76 mostly hypothetical compounds are given. These tables are not meant to give an exhaustive list, but a representative overview for identifying and extrapolating rough trends.

There was no assessment of the surface charges, as we found it to be very constant for fluorine atoms and of quite little significance. All reaction energies (LA, PD, CuD, FIA) are given in kJ mol⁻¹; HOMO and LUMO-HOMO gap are given in eV.

Table 28: Calculated data for [M(CF₃)₄]⁻.

M	sym.	LA	PD	CuD	FIA	HOMO	LUMO-HOMO
B [a]	T	520	-1132	-372	554	-3.532	9.078
Al	T	449	-1203	-443	573	-3.016	6.986
Ga	T	422	-1230	-470	471	-2.918	7.262
Cu [a]	D_{2d}	380	-1272	-512	374	-2.362	3.099
Ag [a]	D_{2d}	390	-1262	-503	367	-2.403	3.695
Au [a]	D_{2d}	420	-1232	-472	383	-2.547	4.667

[a] Taken from Table 25.

Table 29: Calculated data for [M(hfip)ₙ]⁻.

M	n	sym.	LA	PD	CuD	FIA	HOMO	LUMO-HOMO
B	4	S_4	315	-1164	-462	434	-3.383	7.072
Al	4	S_4	392	-1087	-385	535	-3.774	6.716
Ga	4	S_4	392	-1088	-386	504	-3.769	5.778
Si	5	C_1	302	-1177	-475	429	-3.820	6.447
P	6	C_1	246	-1233	-531	435	-4.016	5.390
PO	4	C_2	219	-1260	-558	338	-3.105	5.428
Sb	6	C_i	342	-1137	-435	522	-4.403	3.445
Bi	6	S_6	342	-1137	-435	493	-4.404	1.604
Ti	5	C_1	332	-1148	-446	458	-4.016	3.536
Zr	5	C_1	358	-1121	-419	486	-4.022	4.929
Hf	5	C_1	371	-1108	-406	499	-4.083	5.500
Nb	6	C_i	318	-1161	-459	496	-4.209	3.783
Ta	6	C_i	339	-1141	-439	511	-4.191	4.442
Fe	4	C_2	290	-1189	-487	451	-1.875[a]	0.730[a]
Cu	4	C_{4h}	405	-1074	-372	447	-3.326	1.488
Ag	4	C_{4h}	407	-1073	-371	456	-3.451	1.712
Au	4	C_{4h}	441	-1038	-336	498	-3.274	2.225

[a] Open-shell system; the HOMO is replaced by the SOMO.

Table 30: Calculated data for [M(dfc⁶)ₙ]⁻.

M	n	sym.	LA	PD	CuD	FIA	HOMO	LUMO-HOMO
Al	4	S_4	688	-1054	-369	559	-4.012	5.174
Sb	6	S_6	673	-1069	-363	556	-4.684	3.316

Nb	6	S_6	611	-1131	-445	489	-4.545	3.801
Ta	6	S_6	631	-1111	-426	508	-4.543	4.438
Au	4	C_4	717	-1024	-339	517	-3.635	2.207

Table 31: Calculated data for [M(pftb)$_n$]$^-$.

M	n	sym.	LA	PD	CuD	FIA	HOMO	LUMO-HOMO
Be	3	C_1	301	-1122	-436	473	-3.597	7.009
Al[59]	4	S_4	342	-1081	-395	537	-4.100	6.747
Ga	4	S_4	310	-1113	-428	499	-4.184	5.758
PO	4	C_1	80	-1343	-657	327	-2.954	4.293
Sb	6	C_1	187	-1235	-550	541	-4.729	1.374
Ti	5	C_1	109	-1313	-628	425	-4.172	2.913
Zr	5	C_1	217	-1205	-520	483	-4.299	4.817
Hf	5	C_1	235	-1188	-502	493	-4.289	5.403
Nb	6	S_6	167	-1256	-571	530	-4.669	3.187
Ta	6	S_6	188	-1235	-549	550	-4.701	3.860
ReO$_2$	4	C_1	225	-1198	-512	342	-4.011	2.053
Zn	3	C_1	265	-1122	-472	387	-3.191	5.470

Table 32: Calculated data for [M(pfn^5)$_n$]$^-$.

M	n	sym.	LA	PD	CuD	FIA	HOMO	LUMO-HOMO
Al	4	S_4	632	-1006	-334	591	-5.295	8.038
Sb	6	C_3	607	-1032	-359	614	-5.811	3.266
Nb	6	C_3	564	-1074	-402	587	-5.826	3.627
Au	4	C_4	559	-1080	-407	496	-4.674	1.980

Table 33: Calculated data for M(bpfipm)$_n$]$^-$.

M	n	sym.	LA	PD	CuD	FIA	HOMO	LUMO-HOMO
Be	3	C_1	609	-1091	-389	504	-3.528	5.214
Al	4	C_1	507	-1192	-491	544	-3.828	5.411
Ni	3	C_1	590	-1109	-408	447	-1.229	0.228
Zn	3	C_1	571	-1129	-427	421	-3.264	4.591
Au	4	C_4	607	-1092	-391	493	-3.714	2.109

Table 34: Calculated data for [M(hftb)$_n$]$^-$.

M	N	sym.	LA	PD	CuD	FIA	HOMO	LUMO-HOMO
Al	4	S_4	385	-1100	-398	511	-3.363	6.715
Ga	4	S_4	349	-1135	-433	431	-3.438	5.770
P	6	S_6	49	-1435	-734	409	-3.615	3.348
PO	4	C_1	110	-1374	-673	273	-2.336	4.766
Sb	6	S_6	266	-1218	-516	484	-4.070	2.346
Ti	5	C_1	203	-1282	-580	379	-3.492	3.150
Nb	6	S_6	230	-1255	-553	441	-3.771	3.370
Ta	6	S_6	254	-1230	-529	461	-3.802	4.042
Au	6	C_i	306	-1179	-477	394	-3.886	0.637

Table 35: Calculated data for miscellaneous anions [ML$_n$]$^-$.

L	M	n	sym.	LA	PD	CuD	FIA	HOMO	LUMO-HOMO
mpfc^6	Be	3	C_3	581	-1168	-471	437	-3.292	4.763
mpfc^6	Al	4	S_4	689	-1060	-363	561	-3.825	5.021
pfad	Be	3	C_3	510	-1086	-412	502	-4.112	4.584

pfad	Al	4	S_4	577	-1019	-345	604	-4.581	4.768
pfad	Zn	3	C_3	448	-1148	-474	394	-3.880	4.241
pfbc[7]	Al	4	S_4	574	-1038	-360	570	-4.243	5.243
pfbc[8]	Al	4	S_4	531	-1028	-351	586	-4.392	5.562

Table 36: Calculated data for $[M(pfp)_n]^-$.

M	n	sym.	HOMO	LUMO-HOMO
B	2	S_4	-3.048	6.394
Al	2	S_4	-3.442	6.339
P	3	D_3	-4.137	6.056
PO	2	C_2	-2.645	5.023
As	3	D_3	-4.306	4.295
Sb	3	D_3	-4.464	3.323
Nb	3	D_3	-4.398	3.512
Ag	2	D_2	-3.178	1.121
Re	4	C_1	-4.104	0.926
Au	2	D_2	-3.080	1.812
Au	3	D_3	-4.060	1.125

3.3.2. Results

A superficial view onto Table 28 to Table 36 shows that in a homologous series of elements with the same number and type of ligands attached, the HOMO of the WCA generally tends to decrease with increasing atomic number. For the HOMO-LUMO gap, the picture is inconsistent. Ligand affinity is highly dependent on electronic interactions between ligand and central atom, but also on their sizes.

For P and As, pftb is too large a ligand for six of it to fit around the central atom; the sixth ligand stays loosely in an outer coordination sphere far from the central atom. Even for $[P(hftb)_6]^-$, the LA is quite small, meaning decomposition by ligand abstraction is easily achieved.

Super-Lewis acidic compounds are the ones with a FIA higher than that of SbF_5 (489).[59,72] Many of the investigated compounds surpass this value. The highest known FIA up to date is 716 for $CB_{11}F_{11}$.[73] The synthesis of the Lewis acids, however, is not as trivial as that of their corresponding WCAs because of their intrinsic high reactivity. In the case of hfip, hftb and pftb, the FIA is always higher than the LA; in the case of dfc[6] (and also mpfc[6], where only two compounds were calculated), the case is reversed, meaning that this ligand is able to overcome the energetic disadvantage stemming from its sterical hindrance and form a stronger bond to the central atoms than F^- would do. All other ligands show mixed behavior in this respect, depending on the central atom.

For hfip, hftb and pfp, with P as center, we tested the replacement of two ligands (one in the case of pfp) against a single oxygen atom. In all three cases, the HOMO would be lowered, making the anion easier to oxidize. In the first two cases (where it is defined), the FIA is lowered, also an undesirable trait. Furthermore, the coordination ability is expected to increase because of the "naked" oxygen.

For the bidentate pfp complexes in Table 36, it is unclear how to determine FIA, PD, CuD and LA values. The removal of an entire pfp ligand is thought to be much harder than that of any unidentate ligand because of the chelate effect. The charge of the complex would increase to +1 when a ligand is removed, meaning its PD/ CuD/ LA values would not be comparable to the ones of the neutral Lewis acids. The removal of one single coordination site (i.e. bond breaking between one oxygen and the central atom) would result in a transition state which is non-trivial to find in the gas phase. Similar problems occur when trying to find a reaction to determine the FIA.

Table 37: Calculated data for [AlL$_4$]⁻ anions with miscellaneous ligands L. All energies are given in kJ mol⁻¹ except for HOMO and HOMO-LUMO, which are given in eV).

L	LA	PD	CuD	FIA	HOMO	HOMO-LUMO
hftb	385	-1100	-398	511	-3.363	6.715
hfip	392	-1082	-376	535	-3.774	6.716
pftb	342	-1081	-395	537	-4.100	6.747
bpfipm	507	-1192	-491	544	-3.828	5.411
dfc[6]	688	-1054	-369	559	-4.012	5.174
mpfc[6]	689	-1060	-363	561	-3.825	5.021
pfbc[8]	531	-1028	-351	586	-4.392	5.562
CF$_3$	449	-1203	-443	573	-3.016	6.986
pfbc[7]	531	-1038	-360	570	-4.243	5.243
pfn[5]	632	-1006	-334	591	-5.295	8.038
pfad	577	-1019	-345	604	-4.581	4.768

Table 37 shows aluminate complexes with all of the different ligands, sorted by increasing FIA. The FIA values are on average 561±28 kJ mol⁻¹. Since there is no significant sterical hindrance to be expected for most of the AlL$_3$ and [AlL$_3$F]⁻ species, the FIA could give a pure, unadulterated view on the electron-withdrawing influence different ligands have on aluminium and probably other central atoms as well, if the coordination numbers are the same. In this respect, the pfad ligand has the strongest effect (due to its high electron-withdrawing effect), hftb the weakest. The PD and CuD values are quite close, too: -1053±32 and -367±22 kJ mol⁻¹, respectively, when bpfipm and CF$_3$ are excluded because of their much worse values. For bpfipm, this may be due to sterical overcrowding; for CF_3, because of the lower stability of the Al-C bond.

The ligand affinity has a larger range; it is indicative not only of the affinity of the Lewis acid to the ligand, including sterical effects, but also of the stability of the free ligand L^-. Thus, we may say that free $[mpfc^6]^-$ and $[dfc^6]^-$ possess a rather low stability in the gas phase, free $[pftb]^-$ a rather high one.

The introduction of the pfn^5 ligand will give highest resilience against either oxidation or reduction. The latter, measured by the HOMO-LUMO gap, is the highest by far for all aluminates. Unfortunately, sterical shielding of the – rather strongly basic – oxygen atoms is not very high. Also, the shortest O-C bond of all investigated WCAs is found in $[Al(pfn^5)_4]^-$ (see Table 7), which could lead to think that the C atom tends to form a double bond with cleavage of a fluoroacyl molecule, i.e. the process

(I) $[Al(pfn^5)_4]^- \rightarrow [Al(pfn^5)_3F]^- + (CF_3)_3C\text{-}C(O)F$

The energy of this reaction can be calculated by taking the LA, subtracting the FIA, adding the experimental FIA of $[COF_3]^-$ (209 kJ mol^{-1}), and subtracting the energy of the following reaction (43 kJ mol^{-1}, calculated with MP2/TZVPP):

(II) $COF_2 + [pfn^5]^- \rightarrow (CF_3)_3C\text{-}C(O)F + [COF_3]^-$

Then, we obtain quite a high decomposition energy of +293 kJ mol^{-1} for reaction (I), meaning the WCA would be stable under normal circumstances. If the reaction energies for the decomposition of pfn^5 WCAs with Sb, Nb and Au as central atoms are calculated likewise, values of 244, 215, and 315 kJ mol^{-1}, respectively, are obtained. Of course, an entropic contribution stemming from the evaporation of the fluoroacyl species, especially at elevated temperatures, could aid in decomposition.

There's a multitude of possible reactions to introduce a ligand (OR) to form either a WCA or the corresponding Lewis acid. First, there are addition reactions:

(III) $R_2CO + M\text{-}R' \rightarrow M\text{-}OCR_2R'$

where M-R' is a metal compound with a bond that facilitates addition of a C atom, for example Li[AlH$_4$], Al(CH$_3$)$_3$, LiH; R_2CO is a ketone like perfluorocyclohexanone or perfluorobis(isopropyl)ketone, which then forms a dfc^6, an $mdfc^6$ or a bpfipm ligand. The reaction to form a methyl-substituted bpfipm ligand is not advisable, since sterical overcrowding would be expected, restricting its usability.

A special case of this would be the employment of Li-tert.butyl as a hydride source:[74]

(IV) $R_2CO + Li\text{-}C(CH_3)_3 \rightarrow Li\text{-}OCR_2H + CH_2=C(CH_3)_2$

Preliminary results show that perfluorocyclohexanone reacts with either tert.butyl-lithium or hexyllithium at -78 °C in pentane or hexane to form Li-dfc^6 as evidenced by both IR and NMR spectroscopy.

Second, there's the replacement reaction:

(V) R-OH + M-R' → M-OR + H-R'

where R-OH is an alcohol with an acidic proton and the leaving compound H-R' needs to have a lower boiling point than R-OH and can therefore be evaporated from the system. R' will mainly be H or CH$_3$. Ligands introduced this way may be pfp, hfip, pfbc^7, pfbc^8, pfad, hftb, pftb.

Third, there's the metathesis of a ligand compound M-OR which has to be prepared beforehand, either with the routes shown above or, if applicable, via simple titration with a base in an aqueous system:

(VI) M-OR + M'-X → M-X + M'-OR

where M-X needs to have a higher lattice energy than either M-OR or M'-X. X$^-$ will typically be a small ion like chloride.

Fourth, there's the "PERFECT" process, where a non- or partially fluorinated alcohol is first esterified, then perfluorinated. This ester can then be thermally cleaved to obtain a perfluorinated fluoroacyl compound, from which, for example, an alcohol can be obtained. The procedure was described in detail elsewhere.[75] The pfn^5 alcohol, which is yet unknown, could likely be obtained this way.

The Lewis acid Au(pfn^5)$_3$ is unstable, as, during the gas phase optimization, one ligand gets spontaneously decomposed to form fluoride and fluoroacyl, which both stay coordinated (see Figure 15). However, the assumption we made for the sake of simplicity was that all Lewis acids would be in their monomeric form in the gas phase, which needs not at all to be true; to increase stability, di- or oligomeric forms may be realized as well. In any case, cleavage of the pfn^5 ligand is a possibility that needs to be taken into consideration. For example, in order to rule out any Lewis acidity, it could prove advantageous not to use Au$_2$Cl$_6$, but [AuCl$_4$]$^-$ as a starting material for ligand metathesis according to route (VI) if [Au(pfn^5)$_4$]$^-$ is the desired product.

3.3.3. Conclusion

By calculating key electronic properties beforehand, valuable clues can be gathered which may aid in the directed synthesis of compounds with desired traits. These calculations are inexpensive and can be performed in large quantities at one time, given a computational cluster with a sufficient number of nodes. For a given countercation, suitable anionic candidates of interest may then be chosen for synthesis.

If cost and toxicity are of concern, [Cu(hfip)$_4$]$^-$ (Table 29) could be a valuable target for synthesis, since it is of similar calculated quality as [Al(hfip)$_4$]$^-$, but would facilitate X-ray analysis due to the presence of a heavy atom. However, its small HOMO-LUMO gap would offer little resilience to reduction.

The [Au(dfc^6)$_4$]$^-$ WCA would make one of the best WCAs overall (see Table 30) and arguably the best with commercially available starting materials. It owns the highest LA value as well as excellent CuD and PD values; sterical shielding against electrophilic attack at the oxygen atoms is expected to be of the same order as that of hfip. The cost, however, of the basic keton is almost 100 times the hfip alcohol per mol (see Table 27), making the decision for gold as the central atom less significant.

If a means of fluorination is available, the pfn^5 ligand and ultimately [Al(pfn^5)$_4$]$^-$ could be synthesized, the WCA with the highest PD, the highest CuD, and with FIA and LA among the top five; also, inferred from its HOMO and LUMO, it would be outstandingly resilient against reduction as well as oxidation (see Table 32). Therefore, it could be called the "all-purpose WCA".

4. Ionic Liquids

4.1. General Introduction

In general, an ionic liquid (IL) is any ionic compound that is liquid below 100 °C. So-called "room temperature ionic liquids" (RTILs) are of the highest technical interest. The earliest reported example from almost 100 years ago is a protic RTIL, [C$_2$NH$_3$][NO$_3$].[76] The starting point of modern-day RTIL chemistry was probably the design and investigation of 1:1 salts with delocalized charges, which are often able to avoid crystallization altogether and form glasses far below room temperature instead.[77,78,79]

ILs are unique in their properties in that they are non-volatile and recyclable;[80,81] thus, they are frequently regarded as being potentially tunable materials which could be designed specifically for particular applications like solvents, lubricants, pump oils, phase change media, propellants, and many others.[82,83,84] The large number of salts that have the potential to form ILs and the resulting range of physical properties support this idea.[85,86] However, this also means that the search of a compound with a specific desired property is non-trivial. Therefore, the development of simple methods for the prediction of such properties is a task of major theoretical and technical importance.[87,88,89,90,91,92,93,94,95,96,97,98,99]

4.1.1. The Near-Ordering of Ionic Liquids: Servants of two Masters

Ionic liquids are "servants of two masters", which means that they are governed by the two intermolecular forces named after Coulomb and van der Waals (see Figure 58).[100] Therefore, their physical behavior is a mixture between classical, inorganic salts which are held together mainly by electrostatic forces, and organic compounds whose near-ordering is mostly a consequence of dispersive interactions. In this respect, they are closely related to zwitterions; however, unlike them, sites with positive and negative charges are not held together by a covalent bond, resulting in additional freedom of motion which makes their melting points unusually low.

There were several typical criteria identified for ions to form ionic liquids, among them:[80]
- avoidance of point charges, i.e. a charge that's delocalized over σ- or π-bonds, as in, for example, [PF$_6$]$^-$, [C$_n$MIm]$^+$, or [C$_n$SO$_3$]$^-$;
- low symmetry, as in, for example, [C$_n$MIm]$^+$ with n > 1, to hinder formation of a regular lattice; and

- avoidance of branching, as branched alkyl chains always lead to higher melting points (this is the opposite behavior as observed in alkanes, for example).

Some sort of nanoscale structuring of ILs, also called micro-biphasic behavior for the presence of both polar and non-polar domains, is broadly agreed upon;[101] it was inferred from both simulation and experiment (diffraction, Raman scattering and optical Kerr effect).[102,103,104,105,106,107,108,109,110,111,112,113] The size of these nanodomains is dependent on the length of the alkyl chains and was experimentally found to have a sharp maximum at (or slightly below) the vitrification point.[102a] In a similar vein, there were MD simulations of how water behaves in a bulk IL.[114,115,116] It was found that it forms aggregates, i.e. clusters which extend and connect to form porous networks as the concentration rises. These water domains reorient the surrounding IL molecules to have their hydrophilic sites point towards them.[117,118,119] Thus, mixing in water dramatically changes the internal structure, while energetic changes are probably a minor concern, apart from solvation energies which are neglectable at small moisture contents. This might be compared to the structure of soapy foam where a very small amount of aqueous soap radically alters the properties of the air which it encloses. Thus, it seems that physical quantities of ILs which change rapidly with water content, like melting and vitrification point, conductivity, viscosity, and polarity, are more sensitive to changes in the internal structure.[120,121] Wherever in the following study the water content was deemed critical, we put an emphasis on the different reliability of measured data containing different levels of moisture.

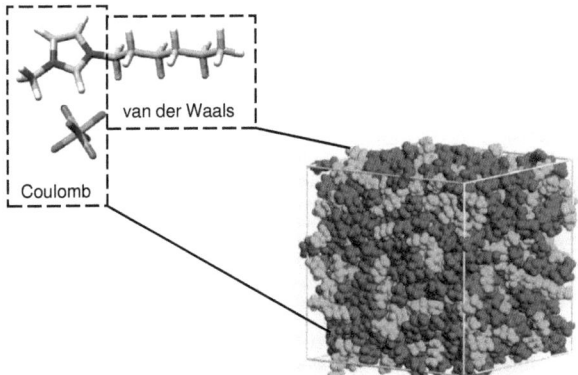

Figure 58: Different domains within the ionic liquid [C_6Mlm][PF_6] as resulting from an MD simulation; within the cube, polar parts are tinted dark, non-polar ones light gray.[122,123] Picture taken from A. Pádua.[124]

4.2. The Molecular Volume[125]

In the context of volume-based thermodynamics (VBT), the molecular volume (V_m) is the basic physical observable used to describe physical properties of ILs. Hitherto, V_m was used in VBT to calculate the lattice energy and the entropy of solids.[126,127,128,129,130] Recent studies connected quantities like viscosity, density, electrical conductivity, and melting point of ILs to V_m (see Figure 59).[87,88,89,90] These relationships may be the basis for simple tools to predict physical properties of yet unknown ILs prior to their synthesis.

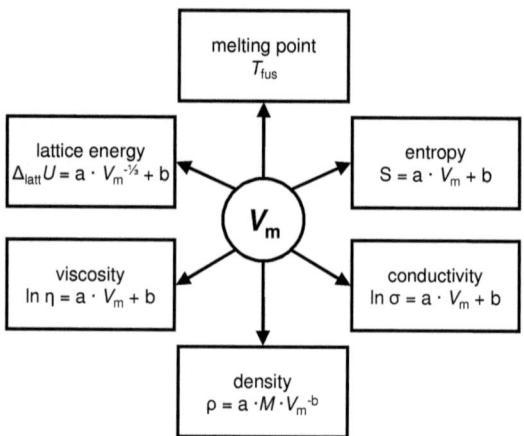

Figure 59: The central role of V_m for the prediction of physical properties of ionic liquids. Coefficients a and b are empirical constants of best fit which differ between the properties. M is the molecular mass.

By definition, the molecular volume V_m refers to the solid state. For an 1:1 salt [A]$^+$[X]$^-$, it is defined as the sum of the ionic volumes V_{ion}([A]$^+$) and V_{ion}([X]$^-$). From the experimental unit cell dimensions of X-ray crystal structures, V_m and the ionic volumes are calculated by eqn. (7):

$$(7) \quad V_m = V_{ion}(A^+) + V_{ion}(X^-) = \frac{V_{cell}(A^+X^-)}{Z}$$

V_{cell} is the volume of the unit cell of the crystal and Z is the number of formula units in the cell. If the ionic volume of one of the ions is accurately known (e.g. halide anions or alkaline metal cations), it may be used as a reference ion to determine the ionic volume of the other ion in the structure.[129,89] This procedure was previously used to derive the experimental ionic volumes for a number of anions and cations commonly used in ILs. Thus, in order to obtain V_m of the ionic liquid in question from experimental studies, suitable crystal structure data is needed.[127] At this point, it should be noted that the temperature of the X-ray measurement is not considered in any of the currently tabulated[129,87,89] V_m and V_{ion} values and, since typical X-ray data collections are performed at temperatures between 90 and 298 K, a small scattering is expected due to the neglect of thermal expansion. Indeed, typical error bars for ionic volumes are in the order of 0.005 to 0.02 nm³. Despite this deficiency, the tabulated X-ray volumes in the solid state proved to be very valuable as an ordering principle for physical properties of liquid ILs, as well as for the prediction of lattice energies of solids as needed in extended BFH cycles.[90,87]

Rebelo et al. used a combination of vdW volumes, density measurements and group contribution methods to obtain the molar volume V_{molar} of ILs in the liquid state (which is related to V_m).[131,132,133] This is an *a posteriori* approach depending on experimental measurements, though. *A priori* (as necessary for the prediction of physical IL properties preceding synthesis), V_m can be approximated by several methods that we found to be lacking in one way or another. Atomic and/ or group contribution methods were widely used, but they are almost always restricted in their choice of possible molecules.[134,135,136,137] In addition, tabulated volumes of ions in solution are dependent on solvent and concentration (which may be idealized as infinite dilution).[138,139] For this study, we chose Hofmann's atomic volumes[134] for comparison; however, we have already shown that the volumes obtained by this approach are less correlated with the physicochemical properties than V_m obtained by X-ray analysis.[87] Gavezzotti integrated the space occupied by the molecule, bounded by vdW radii.[140] Dixon et al. report on a quantum chemical approach to assess ionic volumes.[141] A B3LYP calculation with double-zeta basis sets was used and V_m was assessed by calculating the volume enclosed within the charge density boundaries up to a limiting value of one electron per nm³; this method is computationally rather expensive and lacks the desired simplicity necessary for predictions.[142] Furthermore, this arbitrary approach disregards the fact that different atoms have different electronegativities and polarizabilities, suggesting that there is no universal cut-off of the electron density.

Markedly, none of the aforementioned methods takes charge into account; however, since the free volume is by definition always added to the anion, cations of the same summary formula are always smaller than anions.[129,87] For example, the experimental ionic volumes of $[NO_2]^+$ and $[NO_2]^-$, as given in ref. [129], are 0.022 and 0.055 nm³, respectively. Consequently, Wherland et al. used a charge-dependent fit for the volumes calculated with a method similar to Gavezzotti's.[143] In the current contribution, we wanted to systematically combine these ideas and improve on them to calculate ionic volumes that quantitatively agree with the solid state X-ray volumes. An emphasis was put on those ions relevant to ionic liquids. Thus, our methodology gives values that may directly be used for the prediction of properties that depend on V_m (see Figure 59).

4.2.1. Background

As a basis to estimate ionic volumes V_{ion}, we selected 20 cations and 20 anions with experimentally well-established X-ray structures that were taken from the literature (see Table 38).

Table 38: Ions, their X-ray volumes and assigned errors [in nm³]. Please also consider the supplementary information of the given references, since in some cases the volumes were only established there.

cation	V_{ion}	error	ref.	anion	V_{ion}	error	ref.
[N₂H₅]⁺	0.030	-/-	89	[CN]⁻	0.050	0.006	129
[HPy]⁺	0.095	-/-	89	[OCN]⁻	0.054	0.002	129
[C(N₂H₃)₃]⁺	0.105	-/-	89	[NO₂]⁻	0.055	0.007	129
[N₁,₁,₁,₁]⁺	0.113	0.013	129	[N₃]⁻	0.060	-/-	89
[P₁,₁,₁,₁]⁺	0.133	-/-	89	[SCN]⁻	0.071	0.003	129
[C₂MIm]⁺	0.156	0.018	87	[ClO₄]⁻	0.082	0.013	129
[C₂(CN)MIm]⁺	0.167	0.013	87	[FSO₃]⁻	0.088	0.004	129
[S₂,₂,₂]⁺	0.177	0.017	87	[Dca]⁻	0.089	0.010	87
[C₄MIm]⁺	0.196	0.021	87	[C₁SO₃]⁻	0.099	-/-	89
[N₁,₁,₁,₄]⁺	0.198	0.013	87	[PF₆]⁻	0.107	-/-	89
[N₂,₂,₂,₂]⁺	0.199	0.016	129	[OTfa]⁻	0.108	-/-	89
[C₅MIm]⁺	0.219	0.015	87	[AsF₆]⁻	0.110	0.007	129
[C₄(CN)MIm]⁺	0.222	0.013	87	[BiF₆]⁻	0.124	0.014	129
[C₄C₁MIm]⁺	0.229	0.012	87	[OTf]⁻	0.129	0.007	141
[C₅MPyr]⁺	0.238	0.018	87	[Tcm]⁻	0.130	0.006	144,145,146
[C₆MIm]⁺	0.242	0.015	87	[AlCl₄]⁻	0.161	0.004	147
[N₂,₂,₂,₅]⁺	0.268	0.016	87	[AlBr₄]⁻	0.198	0.005	129
[S₁,φ,φ]⁺	0.268	0.015	87	[Sb₂F₁₁]⁻	0.227	0.020	129
[C₈MIm]⁺	0.288	0.015	87	[NTf₂]⁻	0.232	0.015	87
[N₄,₄,₄,₄]⁺	0.383	0.008	87	['HF']⁻	0.585	0.009	19,148

The following calculations were performed on each ion:
1. Hofmann's simple incremental method.[134]

2. Gas phase structure optimizations with PM3[149], PM6[150], BP86/SV(P)[4,5,6,7] and BP86/TZVP[12], each followed by applying Gavezzotti's method.[140]

3. The same optimizations as above, but with COSMO applied.[13] For the DFT calculations, the dielectric constant, ε_r, was set to ∞; for PM3 and PM6, it was set to 999.0, the maximum value possible within this specific implementation. COSMO constructs a cavity based on optimized radii, which, for most elements, are the vdW radii multiplied by 1.17; the volume of this cavity then is an estimation of the ionic volume V_{ion}.[15,151,152]

In each case, all sets of calculated volumes, V_c, were found to correlate linearly with the experimental ion volumes listed in Table 38 according to

(8) $V_{ion} = a \cdot V_c + b$,

with different parameters a, b for cations and anions and for each of the methods.

Solvation models like the "Isodensity Surface Polarized Continuum Model" (IPCM) and the "Self-Consistent Isodensity Surface Polarized Continuum Model" (SCI-PCM) provide an isodensity surface of the ion in question which may be integrated to give the enclosed volume and thus also would have been interesting to study V_{ion}.[153] However, these methods are tunable with many variables and each parameter also influences the output volume, which is why we found their exploration to be beyond the scope of the present study.

4.2.2. Results

The results are collected in Table 39 and Table 40; +G stands for "Gavezzotti's method applied", +C for "COSMO applied". The experimental $err_\emptyset$ of the volumes obtained by X-ray analysis is 7.3% for the 16 cations and 5.6% for the 16 anions from our list where an error was defined. For comparison, we also give the unscaled (i.e. a = 1 and b = 0) Hofmann values, as he intended them in the original publication (designated Hofmann$_0$).

Table 39: Correlation parameters for the 20 experimental cationic volumes from Table 38 vs. their calculated counterparts (V_c).

method	a	b [nm³]	r²	err$_\emptyset$ (%)
Hofmann$_0$	1.000	0.000	0.9865	10.8
Hofmann	0.964	-0.007	0.9865	5.4
PM3+G	1.458	-0.019	0.9918	3.8
PM6+G	1.437	-0.016	0.9897	4.4
BP86/SV(P)+G	1.420	-0.013	0.9823	5.3
BP86/TZVP+G	1.411	-0.013	0.9855	5.9
PM3+C	1.070	-0.017	0.9905	3.7

PM6+C	1.062	-0.017	0.9899	3.8
BP86/SV(P)+C	1.063	-0.018	0.9925	3.4
BP86/TZVP+C	1.080	-0.019	0.9925	3.3

Table 40: Correlation parameters for the 20 experimental anionic volumes from Table 38 vs. their calculated counterparts (V_c).

method	a	b [nm³]	r²	err$_\varnothing$ (%)
Hofmann$_0$	1.000	0.000	0.9864	21.2
Hofmann	0.946	0.027	0.9864	7.8
PM3+G	1.423	0.010	0.9918	5.6
PM6+G	1.427	0.011	0.9923	5.1
BP86/SV(P)+G	1.457	0.008	0.9910	5.7
BP86/TZVP+G	1.377	0.013	0.9913	5.9
PM3+C	0.966	0.020	0.9831	14.0
PM6+C	0.964	0.020	0.9826	13.9
BP86/SV(P)+C	1.051	0.002	0.9969	4.4
BP86/TZVP+C	1.031	0.004	0.9967	4.3

The need for a charge-dependent linear fit is obvious: The unscaled Hofmann$_0$ volumes give a remarkably larger err$_\varnothing$ than the scaled Hofmann ones.[154] The difference in electron count leads to a negative y-intercept (b-value) for the cations and a positive value for the anions. The higher slopes (a-values) for Gavezzotti's method are caused by the aforementioned fact that it uses non-scaled vdW radii, while COSMO uses scaled ones. The fit also has the advantage that values for V_{ion}, either experimentally determined or calculated with any method, can be arbitrarily combined to give V_m.

Overall, for every method, the correlation between V_c and V_{ion} is very high, with $r^2 > 0.98$. Hofmann's volumes, either with or without a fit, are the worst methods. The semi-empirical models PM3 and PM6 perform well for the cations (err$_\varnothing$: 3.7 to 4.4%), however, especially the errors of the anion volumes using PM3 or PM6 together with COSMO (err$_\varnothing$: 13.9 to 14.0%) are very large and thus these models are not as well suited as others. For BP86/SV(P) or BP86/TZVP with COSMO, and for almost all cases with Gavezzotti's method applied, the err$_\varnothing$ values of V_{ion} (3.8 to 5.9%), regardless of charge, are smaller than the experimental error of the volumes obtained by X-ray analysis. In general, predictions of anionic volumes are less reliable than those for cations. Intrinsic to the experimental data, relative errors tend to be higher for smaller ions (highest in [N$_2$H$_5$]$^+$ and [ClO$_4$]$^-$), and diminish with size. Overall, we think that the calculated volumes using the best methods from Table 39 and Table 40 are more internally consistent than V_{ion} determined by X-ray analysis, since the latter ones are not corrected for thermal expansion and the calculated volumes always refer to one point of reference (the converged structure optimization) independent of temperature or counter ion.

With these results in mind, ionic and molecular volumes obtained with the best method for both anionic and cationic volumes, BP86/TZVP+COSMO, were used for all the following calculations and predictions (see Figure 60).

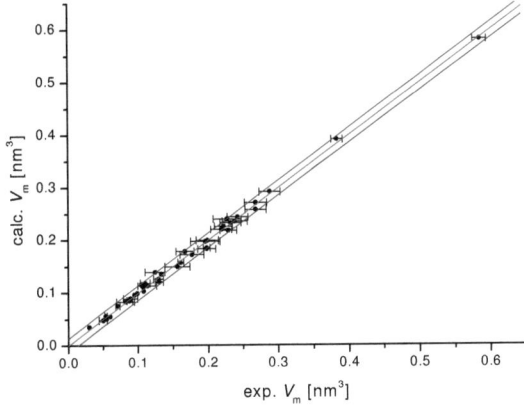

Figure 60: The connection of experimental and calculated (BP86/TZVP+COSMO) molecular volumes, shown for all anions and cations from Table 39 and Table 40 alongside the error bars and 95% prediction bands.

4.2.3. Conclusion

For the calculation of the molecular volume, a geometry optimization with the BP86 functional and either TZVP or SV(P) basis set followed by a COSMO calculation and a charge-dependent fit will give the results with the highest agreement with X-ray data. Second-best, PM3 or PM6 with Gavezzotti's method can give very fast approximations. If only rough estimates are needed, scaled Hofmann volumes might give a good starting point as well. The solid state volume V_m as an *in silico* measure is, as demonstrated later, well-correlated with heat capacity and temperature-dependent density of ionic liquids and allows to approximate physical properties of ILs prior to synthesis. Thus, we propose that for the prediction of physical properties of yet unknown ionic liquids, V_m and its dependent properties should be calculated using our methodology.

4.2.4. Appendix: Additional Computational Details

Gas phase PM3 optimizations were performed with ChemBioOffice 2008, © 1986 – 2007 CambridgeSoft. PM3 optimizations with COSMO as well as all PM6 optimizations were performed with MOPAC2007.[155] The calculations of Gavezzotti's volumes were accomplished using the freeware program steric.[156] All DFT calculations were carried out with the TURBOMOLE program package (version 5.10).[8] The implementations of COSMO are the ones used in the respective programs.

4.3. The Temperature-Dependent Density[125]

The densities of ionic liquids were determined by many different researchers in the past, and temperature-dependent rules were set up. Recently, Coutinho found an equation for the estimation of densities for arbitrary pressures and temperatures.[157] However, he took X-ray volumes as the basis and derived other molecular volumes by group contributions as needed. For testing our best computational method (BP86/TZVP+COSMO) against these results, we chose another, more varied set, including functionalized ILs, and made a different, more precise deduction of the temperature dependency.

The cubic isobaric thermal expansion coefficient α_p, is defined as

(9) $\alpha_p = -\dfrac{1}{\rho}\left(\dfrac{\partial \rho}{\partial T}\right)_p$

and so the density ρ at a given temperature T can be expressed as:

(10) $\ln(\rho/\rho_0) = -\alpha_p \cdot T + \beta$

with $\rho_0 = 1$ g cm^{-1}. Both α_p and β are dependent on the nature of the IL in question. Building on the impressive corpus of experimental studies accomplished by other authors, we compared the quality of eqn. (10) for a representative selection (26 different ionic liquids in 12 references) of test series. Squared correlation coefficients were always > 0.999. We hold the minimum temperature range for the validity of this equation to be 298...323 K, the intersection of all test series. The maximum temperature range is 273...415 K. For the data of ref. [158], analyzing ILs from different manufactures, we took the average densities at the given temperatures.

In our quest for a universal rule, we found the following V_m-based relationships for α_p and β, which are independent from the nature of the IL in question:

(11) $\alpha_p = a \cdot \ln(V_m/V_0) + b$

with a = 0.0001747 K^{-1}, b = 0.0008028 K^{-1}, V_0 = 1 nm^3, and

(12) $\beta = c \cdot \ln\left(\dfrac{M \cdot V_0}{M_0 \cdot V_m}\right) + d$

with c = 1.158, d = −7.413, M = molar mass, M_0 = 1 g mol^{-1}, V_0 = 1 nm^3. This quantity describes the logarithm of the ideal liquid density at 0 K.

Written out, the temperature-dependent, empirical expression for the density ρ of all hitherto investigated ILs is:

(13) $\rho(T) = (V_0/V_m)^{a \cdot T + c} \cdot (M/M_0)^c \cdot \exp(-b \cdot T + d)$

with coefficients a...d and M_0, V_0 as above. Eqn. (13) expands and replaces our old expression for the calculation of IL densities which was valid only at 294 K ($\rho = e \cdot M \cdot (V_m)^{-f}$, with e, f constants).[87]

Densities were calculated according to eqn. (13); then, err$_\varnothing$ was independently calculated for each test series over the entire given temperature range. $[P_{6,6,6,14}][NTf_2]$ and $[N_{1,8,8,8}][NTf_2]$ were excluded[158,165], as they pose exceptions to eqn. (11) (but not to eqn. (12)), maybe because of the formation of liquid crystals.[159] The average error values of all test series for each IL can be found in Table 41. They are comparable with Coutinho's values and rarely exceed 1%. The largest errors occur in the case of water-saturated ILs, functionalized cations or cations with long alkyl chains, which also may form liquid crystal phases.

For eqn. (12), the squared correlation coefficient (r^2) of calculated and experimental data is most excellent (0.9933); for eqn. (11), it is less satisfactory (0.6095), but, even so, the calculated densities are in very good agreement with the experimental ones. It is reasonable to assume that α_p contains information about intermolecular interactions.

Since all coefficients c...f are independent of the specific nature of the IL and, thus, ρ only depends on the molar mass M and calculated volumes V_m, eqn. (13) shows the potential for the *in silico* prediction of densities of hitherto unknown ILs.

From eqn. (13), a simple expression for isobaric expansion at $T_2 = T_1 + \Delta T$ follows, given that the density at any temperature T_1 is known:

(14) $\rho(T_2) = (V_0/V_m)^{a \cdot \Delta T} \cdot \exp(-b \cdot \Delta T) \cdot \rho(T_1)$,

with $V_0 = 1$ nm³ and temperatures in K.

The liquid density, calculated at 0 K using eqn. (13), is highly correlated with V_m according to:

(15) $\dfrac{M}{N_A \cdot \rho(0\ K)} = a \cdot V_m + b$

with N_A = Avogadro's constant, a = 0.9516, b = -0.0056 nm³. Not only is the y-intercept (b) very small, r^2 is also very high (0.9969), assuming that there's a universal direct relationship between the volume a molecule assumes in the disordered, liquid state and in the ordered crystal, even though the size of voids (i.e. unoccupied space) in the lattice assumes discrete values, while in the liquid, they are random.

Table 41: err$_\emptyset$ in % for eqn. (13) and molecular volumes in nm^3 (BP86/TZVP+COSMO), calculated for different test series.

IL	V_m	err$_\emptyset$	ref.	IL	V_m	err$_\emptyset$	ref.
[C$_4$MIm][PF$_6$]	0.307	0.82	158	[C$_4$MIm][NTf$_2$]	0.430	0.09	158
		0.83	166			0.09	165
		0.71	160			0.07	166
		0.80	161			0.09	167
		2.08^a	161			0.27	160
		0.62	162			0.06	161
		0.74	163			1.18^a	161
		1.12	164				
[C$_2$MIm][NTf$_2$]	0.382	0.86	158	[C$_2$MIm][C$_2$SO$_4$]	0.282	0.55	158
		0.75	165			0.63	165
		0.92	161			0.86	166
		2.24^a	161			0.80	161
						0.40	168
[C$_4$MIm][BF$_4$]	0.276	0.40	158	[N$_{1,1,1,4}$][NTf$_2$]	0.416	0.63	165
		0.46	166			0.79	161
		0.42	161			1.40^a	161
[C$_8$MIm][PF$_6$]	0.402	0.18	164	[C$_6$MIm][NTf$_2$]	0.477	0.21	158
[C$_8$MIm][BF$_4$]	0.371	0.36	164	[C$_8$MIm][NTf$_2$]	0.525	1.11	158
[C$_4$Py][BF$_4$]	0.270	0.14	164	[C$_1$MIm][C$_1$SO$_4$]	0.237	0.70	169
[C$_4$MPyr][Fap]	0.547	0.25	158	[C$_6$MIm][Fap]	0.578	0.41	158
[C$_4$Py][Fap]	0.526	0.69	158	[C$_1$(CN)Py][NTf$_2$]	0.381	3.63	158
[C$_4$MIm][OTf]	0.323	1.63	158	[C$_1$(CN)MPyr][NTf$_2$]	0.406	3.63	158
[C$_1$MIm][SCN]	0.201	1.16	158	[C$_{10}$MIm][NTf$_2$]	0.564	1.54	158
[C$_6$MIm]Cl	0.285	1.81	170	[C$_8$MIm]Cl	0.333	2.71	170
[C$_4$MIm][C$_8$SO$_4$]	0.465	0.72	165	[C$_6$MIm][PF$_6$]	0.354	0.05	167

^{a} Water-saturated.

In the meantime, an improved formula to describe α$_p$ was found:

(16) $\alpha_p = a \cdot \ln(r_m/r_0) + b \cdot E_{ZP} + c$

where r_m is the molecular radius (see eqn. (6)), E_{ZP} the sum of the zero-point energies of the separate ions in the gas phase; $r_0 = 1$ nm, a = 0.000514 K^{-1}, b = -0.0000504 mol J^{-1} K^{-1}, c = 0.000841 K^{-1}. Thus, with only one additional descriptor, r^2 is increased from 0.6095 in eqn. (X) to 0.8077, this time including the [NTf$_2$]$^-$ compounds of [P$_{6,6,6,14}$]$^+$ and [N$_{1,8,8,8}$]$^+$ and not having any distinct outliers. Even though b is extremely small, its presence suffices to dramatically increase the correlation coefficient. If we think of the molecule as a harmonic oscillator whose energy is – in a certain region – a linear function of the displacement from the equilibrium state, this displacement could be approximated by the force constant and therefore E_{ZP}. As expected, E_{ZP} has a negative influence on the expansion coefficient, as the negative value of b indicates.

As Table 42 shows, the average errors are usually slightly larger than if eqn. (11) was employed. However, in some cases, it is notably smaller, especially in the cases of [C$_n$MIm]$^+$ with [NTf$_2$]$^-$ and Cl$^-$ (where n = 6, 8), [C$_{10}$MIm][NTf$_2$], and [C$_4$MIm][C$_8$SO$_4$], i.e. compounds with, due to their increased conformational freedom, quite high zero-point energies. The average errors for $[N_{1,8,8,8}][NTf_2]$ and $[P_{6,6,6,14}][NTf_2]$, compounds which could not be treated in a meaningful manner with eqn. (11), are between 1...2%.

Table 42: err$_\varnothing$ in % for eqn. (10) if α_p was calculated according to eqn. (16), alongside molecular radii in nm^3 (BP86/TZVP+COSMO) and zero-point energies in kJ mol^{-1} as calculated for different test series. References are the same as given in Table 41 except for the last two entries.

IL	r_m	E_{ZP}	err$_\varnothing$	IL	r_m	E_{ZP}	err$_\varnothing$
$[C_4MIm][PF_6]$	0.659	626	0.88	$[C_4MIm][NTf_2]$	0.743	711	0.19
			0.89				0.18
			0.77				0.17
			0.85				0.18
			2.14[a]				0.35
			0.67				0.15
			0.80				1.27[a]
			1.18				
$[C_2MIm][NTf_2]$	0.711	565	0.78	$[C_2MIm][C_2SO_4]$	0.646	641	0.62
			0.68				0.70
			0.84				0.93
			2.16[a]				0.88
							0.47
$[C_4MIm][BF_4]$	0.627	614	0.64	$[N_{1,1,1,4}][NTf_2]$	0.734	768	0.82
			0.72				1.00
			0.67				1.60[a]
$[C_8MIm][PF_6]$	0.709	915	0.47	$[C_6MIm][NTf_2]$	0.769	855	0.12
$[C_8MIm][BF_4]$	0.677	904	0.55	$[C_8MIm][NTf_2]$	0.793	1000	0.51
$[C_4Py][BF_4]$	0.624	588	0.05	$[C_1MIm][C_1SO_4]$	0.609	496	1.02
$[C_4MPyr][Fap]$	0.801	959	0.87	$[C_6MIm][Fap]$	0.818	952	0.97
$[C_4Py][Fap]$	0.788	782	1.05	$[C_1(CN)Py][NTf_2]$	0.710	463	3.86
$[C_4MIm][OTf]$	0.672	644	1.66	$[C_1(CN)MPyr][NTf_2]$	0.727	641	3.52
$[C_1MIm][SCN]$	0.573	380	1.58	$[C_{10}MIm][NTf_2]$	0.814	1146	0.65
$[C_6MIm]Cl$	0.601	720	0.55	$[C_8MIm]Cl$	0.625	866	1.00
$[C_4MIm][C_8SO_4]$	0.761	1221	0.19	$[C_6MIm][PF_6]$	0.685	770	0.37
$[N_{1,8,8,8}][NTf_2]$[165]	0.906	2044	1.18	$[P_{6,6,6,14}][NTf_2]$[158]	0.956	2420	1.91

[a] Water-saturated.

4.4. The Integrated Heat Capacity[125]

Because heat capacity (C_p) is dependent on entropy, which, in turn, is related to the molecular volume[127,128], we also explored the possibility of a direct correlation between V_m and C_p. Heat capacity data of ILs measured at different temperatures are available, but the functional for their temperature dependency contains up to 6 coefficients.[171,172] Also, data for the same IL but from different test series tends to possess very dissimilar temperature behavior, mostly caused by varying water content.[173,174] Accordingly, Kabo's correlation scheme shows quite limited applicability.[171] So we refrained from seeking a general rule as for densities and restricted ourselves to two distinct temperatures (298 and 323 K). We calculated V_m of 34 ionic liquids for which the integrated heat capacities were taken from the NIST IL database.[175] For many of these compounds, there are too few, if any, crystal structures to gain a meaningful experimental V_m, so we calculated them using our most reliable method (BP86/TZVP+COSMO). We divided the ionic liquids into two groups, depending on moisture. For set 1 with low (< 500 ppm) water content, see Table 43. Set 2 comprises all ILs of set 1 and, additionally, the ILs listed in Table 44. If more than one data point was available for a specific IL, we took the one with the lowest water content or highest purity, except for [C₄MIm][PF₆], for which we took the mean value of the data in ref. [160] and [176].

Table 43: The ILs of set 1 with V_m [nm³] (BP86/TZVP+COSMO), experimental and calculated C_p^T [J mol⁻¹ K⁻¹] at 298 and 323 K. References are given for the experimental values.

IL	V_m	C_p^{298}		C_p^{323}		ref.
		exp.	calc.	exp.	calc.	
[C₄MIm][NTf₂]	0.430	536.3	550	543.9	572	176
[(C₄)₂Nic][NTf₂]	0.557	707	698	727	723	177
[C₄C₁Py][NTf₂]	0.448	622	571	641	594	177
[C₆C₁Py][NTf₂]	0.495	624	626	644	650	177
[C₄C₁Py][BF₄]	0.294	405	390	421	410	177
[C₆Mim][NTf₂]	0.477	629.2	605	643.2	628	178
[C₂Mim][NTf₂]	0.382	524.3	494	532.2	516	176
[C₆(C₁)₂Py][NTf₂]	0.520	620	655	665	679	177
[C₆C₁DmaPy][NTf₂]	0.565	725	707	764	732	177
[C₆C₃(C₂)₂Py][NTf₂]	0.633	766	788	799	814	177
[C₆DmaPy][NTf₂]	0.535	628	673	650	697	177
[C₂C₁Py][C₂SO₄]	0.301	389	399	402	418	177
[C₆Py][NTf₂]	0.472	612	599	632	622	177
[C₃MPyr][NTf₂]	0.422	554	541	570.5	563	179
[C₄MIm]Br	0.244	296	332	359	351	180
[C₄MIm][BF₄]	0.276	364.7	369	379.0	389	174
[C₂MIm][BF₄]	0.228	308.1	313	316.8	332	179

Table 44: The additional ILs of set 2 with V_m [nm³] (BP86/TZVP+COSMO), experimental and calculated C_p^T [J mol⁻¹ K⁻¹] at 298 and 323 K. References are given for the experimental values.

IL	V_m	C_p^{298}		C_p^{323}		ref.
		exp.	calc.	exp.	calc.	
[nonafluoro-C₆MIm][NTf₂]	0.553	725	693	752	718	177
[(C₂)₂Nic][C₂SO₄]	0.363	513	472	530	493	177
[C₃C₁MIm][NTf₂]	0.430	554.5	550	558.7	572	176
[C₆C₁MIm][NTf₂]	0.498	686	629	705	653	177
[C₈MIm]Br	0.339	392	443	408	464	177
[C₄MIm]Cl	0.238	322.7	325	333.7	344	176
ECOENG 41M	0.423	643	542	652	564	177
[C₆MIm]Br	0.291	344	387	357	407	177
[C₈MIm][BF₄]	0.371	506	480	526	501	177
[C₄MIm][Dca]	0.286	364.6	381	370	400	176
[N₄,₄,₄,₄][Doc]	0.953	1325	1161	1385	1194	177
[C₄MIm][PF₆]	0.307	402.6	406	413.5	426	160,176
[C₆MIm][BF₄]	0.322	416	424	433	444	177
[C₈MIm][NTf₂]	0.525	654	661	677	685	177
[C₄C₁MIm][PF₆]	0.329	433.6	431	449.1	452	176
[C₄MIm][OTf]	0.323	417.2	424	423.1	445	176
[C₂MIm][C₂SO₄]	0.282	378	377	383.9	396	181

The data show a linear relationship between V_m and C_p given by the following empirical equation:

(17) $C_p = a \cdot V_m + b$

The correlation parameters can be found in Table 45. The fit was derived from set 1 only and then tested against set 2. Surprisingly, although this is a very simple relationship, the average errors have about the same order of magnitude as the assigned uncertainty in the experimental C_p determinations. In set 2, the average error rises only by 0.7 to 1.2% (see Figure 61). Interestingly the 323 K data set (which presumably, due to the higher temperature, has lower water content) has smaller average errors than the 298 K data. We found two compounds to pose outliers with 19...20% error: [C₆C₁Py]Br and ECOENG 500.[177] The former has unknown H₂O content; the latter (1044 ppm H₂O) may form mesophases due to its long alkyl chain (tridecyl), thereby influencing C_p. Apart from these exceptions, the *in silico* estimation of the heat capacity of an IL at both temperatures, can be expected to lie close to the experimental value and thus be straight forward to be used for predictions.

For comparison, the average errors for scaled Hofmann volumes were also calculated. The errors we found are always > 5%; especially at 323 K, they are remarkably larger than the errors for the BP86/TZVP+COSMO volumes.

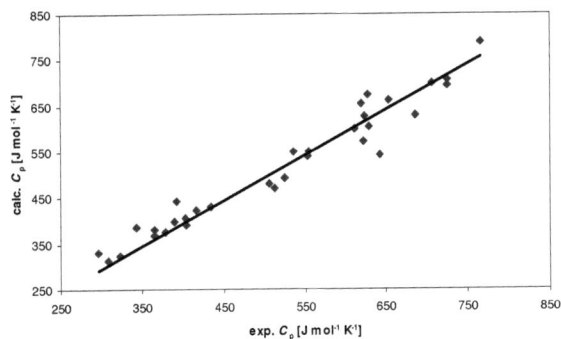

Figure 61: The relationship of experimental and calculated C_p at 298 K for set 2. [N$_{4,4,4,4}$][Doc] was left out due to its extraordinarily high C_p (exp.: 1325 J mol^{-1} K^{-1}).

Table 45: Correlation constants a [in J mol^{-1} K^{-1} nm^{-3}] and b [in J mol^{-1} K^{-1}] in eqn. (17) for the dependency of C_p from V_m (BP86/TZVP+COSMO).

data set	T [K]	a	b	r²	err$_\varnothing$ (%)	err$_\varnothing$ (%)[a]
1	298	1169	47.0	0.9693	3.9	5.5
1	323	1189	61.0	0.9765	3.2	5.9
2	298	1169	47.0	0.9682	4.8	6.0
2	323	1189	61.0	0.9703	4.6	6.8

[a] Using scaled Hofmann volumes. The best-fit coefficients a and b to calculate this error are derived from the dataset with low water content and are 1244 and 27.8 at 298 K and 1226 and 42.8 at 323 K, with units as above.

4.5. The Temperature-Dependent Liquid Entropy[182]

4.5.1. Background

For only about a dozen ionic liquids, there is temperature-dependent entropy data available from literature. For its determination, heat capacity needs to be measured with adiabatic calorimetry, which is extremely tedious and may take up to several months because of the slow tempering of ionic liquids. Also, for ILs that do not crystallize, the entropy cannot be determined.

In the context of VBT, there were several connections made between the molecular volume and the entropy of solids and liquids; however, no parametrization specifically for ILs exists.[127,128] In search for such a universal correlation, we tried to employ COSMO-RS which gives the chemical potential in the gas phase (μ_g^{CT}) as well as in the liquid (μ_l^{CT}).[15] Our test set consists of [C#MIm][NTf$_2$], with # = 2, 4, 6, 8; and [C$_4$MIm][A], with A = NO$_3$, OTfa, OTos, PF$_6$. It turned out to be best to calculate these figures for the paired ions (for gas phase structures see section 2.2.2). Since the difference of the chemical potential in the gas phase and the liquid is proportional to the logarithm of the vapor pressure, a measure which COSMO-RS predicts fairly well,[183] we assumed that $\mu_g^{CT} - \mu_l^{CT}$ has the correct size. For the calculation of μ_g, however, we think it's more advisable to use the output of the statistical thermodynamics calculated with freeh (as implemented in the TURBOMOLE suite of programs), which we will denote as μ_g^{freeh}. Surprisingly, it is quite different from μ_g^{CT} for any compound. Even more surprisingly, μ_l^{CT} rises with temperature, which is physically meaningless.

In combination, we therefore corrected μ_l by the following equation:

(18) $\mu_l^* = \mu_g^{freeh} - (\mu_g^{CT} - \mu_l^{CT})$

Then, the liquid entropy can be obtained by partial differentiation of μ_l^* with respect to T:

(19) $S_l^* = -\dfrac{\partial \mu_l^*}{\partial T}$

4.5.2. Results

We found that, while the correlation of S_l^* and the measured S_l^{exp} over the observed temperatures is quite strong for every IL of the test set ($r^2 \geq 0.9976$), the average error is up to 7.1%, a clear sign that linear combination parts are missing in this equation. Since, for calculations of Gibbs energies, entropy is scaled with T, a slight error in S can introduce quite large errors in ΔG. We also found that S_g^{CT}, as the negative local derivative of μ_g^{CT} with respect to T, is a constant (81.4 J mol^{-1} K^{-1}) for every compound and every temperature, which follows immediately from the empirical equation used in COSMOtherm (the employed implementation of COSMO-RS):[184,14]

(20) $\mu_g^{CT} = E_g - E_{COSMO} - E_{vdW} + \omega_{ring} n_{ring} + \eta_g RT$

Here, E_g and E_{COSMO} are the total energies in the gas phase and in the ideal conductor, respectively; n_{ring} is the number of ring atoms, ω_{ring} a constant. Since the first four addends in this equation are temperature-independent and η_g is constant, the derivative of μ_g^{CT} with respect to T is also constant.

So this term was left out and S_l was written as a linear combination of the calculated entropies:

(21) $S_l^{**} = a \cdot S_g^{freeh} + b \cdot S_l^{CT} + c$

where S_g^{freeh} can be calculated directly with freeh and

(22) $S_l^{CT} = -\dfrac{\partial \mu_l^{CT}}{\partial T}$

Table 46: The connection of measured and calculated entropies. err$_\varnothing$ (1) belongs to eqn. (19), err$_\varnothing$ (2) to eqn. (21).

IL	temperature range	number of data points	r²	err$_\varnothing$ (1)	err$_\varnothing$ (2)	ref.
[C$_2$MIm][NTf$_2$]	260...370	14	0.9998	5.5%	1.6%	185
[C$_4$MIm][NO$_3$]	309.16...370	7	0.9976	7.1%	3.4%	186
[C$_4$MIm][NTf$_2$]	190...370	14	0.9991	2.1%	2.1%	185,31
[C$_4$MIm][OTfa]	190...370	20	0.9991	5.5%	2.1%	171
[C$_4$MIm][OTos]	343.89...470	13	1.0000	3.3%	1.4%	187
[C$_4$MIm][PF$_6$]	200...330	15	0.9993	1.0%	3.1%	163
[C$_6$MIm][NTf$_2$]	190...370	20	0.9989	4.6%	1.1%	188
[C$_8$MIm][NTf$_2$]	263.96...370	12	0.9997	5.8%	1.4%	185

A multidimensional linear fit to the existing data – kindly provided by Y. Paulechka, University of Minsk, Belarus – gave a = 1.059, b = -1.122, c = -90.01 J mol^{-1} K^{-1}. As Table 46 shows, the average absolute error is in general significantly reduced, with the notable exception of [C$_4$MIm][PF$_6$] which had a very low error with eqn. (19) to boot. As expected, r² does not change. Internal rotations (e.g. the rotation of a -CH$_3$ group) are not accounted for in both freeh and COSMO-RS since in freeh, only torsion angles are being varied; in COSMO-RS, the chemical potential is calculated by statistical analysis of interactions in the liquid, leaving no true variation of any degree of freedom. Probably, these circumstances lead to coefficients a and b ≠ 1. Even though the magnitudes of a and b are very similar, we decided not to merge them because of the different calibrations of the quantum chemical methods used to obtain μ_g^{freeh} and μ_l^{CT} and because of the aforementioned fact that the error in entropy needs to be kept at a minimum. Figure 62 is a graphical representation of these findings.

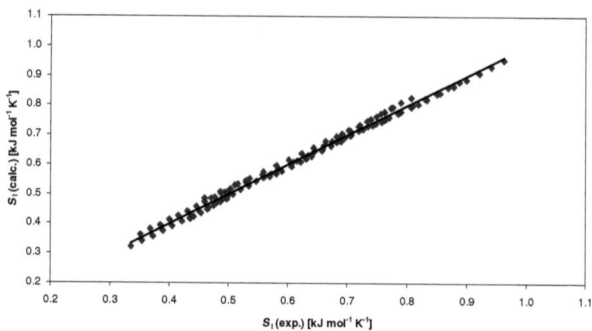

Figure 62: All experimental and calculated (using eqn. (21)) liquid entropies of Table 46.

Subsequently, we tried to find a quicker way, using separate ions instead of ion pairs. The gas phase entropies are the sum of freeh's anionic and cationic entropy, the liquid entropy is calculated from a 1:1 mixture of anion and cation with COSMO-RS. Optimized, the formula reads:

(23) $S_l^{\pm**} = a \cdot S_g^{\pm,freeh} + b \cdot S_l^{\pm,CT} + c$

where a = 1.009, b = -0.1928, c = -275.9 J mol⁻¹ K⁻¹. As Table 47 shows, the errors are in most cases considerably higher than for eqn. (21). The highest error can be found in the rather strongly coordinating [C₄MIm][NO₃], giving a hint of the interactions which are missing if the ion pair is broken up. However, the computational cost is significantly lower than for the ion pairs.

Table 47: The connection of measured and calculated (according to eqn. (23)) entropies.

IL	r^2	err$_\emptyset$	IL	r^2	err$_\emptyset$
[C₂MIm][NTf₂]	0.9972	1.9%	[C₄MIm][OTos]	0.9419	3.8%
[C₄MIm][NO₃]	0.9976	7.9%	[C₄MIm][PF₆]	0.9866	3.3%
[C₄MIm][NTf₂]	0.8931	5.1%	[C₆MIm][NTf₂]	0.9986	4.1%
[C₄MIm][OTfa]	0.9976	3.6%	[C₈MIm][NTf₂]	0.9839	1.6%

To find a simple formula for the standard liquid entropy of all kinds of imidazolium ILs, we then investigated 40 imidazolium ILs for which we calculated S_l^{**} with eqn. (21) as the best approximation, i.e. using COSMO-RS and a frequency calculation with AOFORCE and a vibrational analysis with freeh on the optimized ion pair. At 25 °C, we found S_l^{**} to correlate nicely with $S_g^{0,\pm}$. We also found a good correlation with r_m^3 which is unsurprising since standard entropies in the solid state are a function of V_m and therefore a correlation with the related molecular radius could also be expected.[127,128]

(24) $S_l^{0,**} = a \cdot S_g^{0,\pm} + b$

(25) $S_l^{0,**} = a \cdot r_m^3 + b$

The results are given in Table 48 and Table 49.

Table 48: Calculated entropies (according to eqn. (21); given in kJ mol^{-1} K^{-1}) and molecular radii.

IL	r_m [nm]	$S_g^{0,\pm}$	$S_l^{0,**}$	IL	r_m [nm]	$S_g^{0,\pm}$	$S_l^{0,**}$
[C₂MIm][BF₄]	0.229	0.658	0.370	[C₄MIm][NO₃]	0.233	0.688	0.394
[C₂MIm][C₂SO₄]	0.248	0.748	0.408	[C₄MIm][NTf₂]	0.286	0.980	0.689[a]
[C₂MIm][Dca]	0.233	0.658	0.364	[C₄MIm][OAc]	0.241	0.729	0.410
[C₂MIm][NO₃]	0.221	0.626	0.335	[C₄MIm][OTf]	0.259	0.806	0.486
[C₂MIm][NTf₂]	0.274	0.918	0.629[a]	[C₄MIm][OTfa]	0.251	0.783	0.498[a]
[C₂MIm][OAc]	0.229	0.667	0.353	[C₄MIm][PF₆]	0.253	0.754	0.493[a]
[C₂MIm][PF₆]	0.241	0.692	0.413	[C₄MIm][SCN]	0.240	0.675	0.354
[C₂MIm][SCN]	0.227	0.612	0.295	[C₄MIm][Tcm]	0.257	0.759	0.448
[C₂MIm][Tcm]	0.245	0.697	0.391	[C₄MIm]Br	0.225	0.605	0.342
[C₂MIm]Br	0.213	0.543	0.270	[C₄MIm]Cl	0.221	0.595	0.316
[C₂MIm]Cl	0.209	0.532	0.262	[C₆MIm][NTf₂]	0.296	1.058	0.755[a]
[C₃MIm][Dca]	0.240	0.690	0.384	[C₆MIm][PF₆]	0.264	0.832	0.527
[C₃MIm][NO₃]	0.228	0.657	0.372	[C₈MIm][BF₄]	0.261	0.858	0.541
[C₃MIm][OAc]	0.235	0.699	0.382	[C₈MIm][Dca]	0.265	0.858	0.525
[C₃MIm][PF₆]	0.248	0.724	0.444	[C₈MIm][NTf₂]	0.305	1.118	0.816[a]
[C₃MIm][SCN]	0.234	0.644	0.329	[C₈MIm][OTf]	0.278	0.944	0.601
[C₃MIm][Tcm]	0.251	0.729	0.418	[C₈MIm][PF₆]	0.273	0.892	0.622
[C₃MIm]Cl	0.215	0.564	0.288	[C₈MIm]Cl	0.241	0.733	0.429
[C₄MIm][BF₄]	0.241	0.721	0.431	[φCCCMIm][NTf₂]	0.303	1.069	0.723
[C₄MIm][Dca]	0.245	0.720	0.413	[φCCCMIm][PF₆]	0.270	0.844	0.520

[a] Experimental value.

Table 49: Coefficients used in eqn. (24) and (25).

eqn.	a	b	r^2	err$_\emptyset$
(24)	0.9402	-0.2560 kJ mol^{-1} K^{-1}	0.9783	3.9%
(25)	1.585 kJ mol^{-1} K^{-1} nm^{-3}	0.01409 J mol^{-1} K^{-1}	0.9582	5.1%

No specific outliers were observed. Of course, since we calibrated the coefficients mainly against calculated and not experimental values, the true error would probably be somewhat higher.

For comparison, if we set S_g^0, calculated from ion pairs, into eqn. (24), we get $r^2 = 0.9842$, $err_\varnothing = 2.9\ \%$, but at a much higher computational cost.

4.5.3. Conclusion

We found a simple, VBT-like equation, parametrized especially for imidazolium ionic liquids and only dependent on the molecular radius, to describe the liquid entropy at standard conditions. To include temperature dependency, the more demanding calculation of an ion pair is advisable and the results are typically only 1...2% off from the experimental results in a large temperature range. If single ions are used, the errors roughly triple, but the calculation speed is greatly increased. Future studies may comprise a parametrization for other kinds of ionic liquids as well as inclusion of the solid-state entropy.

4.6. Basic Phase Change Thermodynamics

Born-Fajans-Haber (BFH) cycles find widespread application in many areas of chemistry in order to gain insight into thermodynamic relations, for example between starting materials and products of a chemical reaction, or between different states of matter of the same substance.[189] For ionic liquids, the interest for their state of matter has surged since they were shown to evaporate given sufficient high temperature and low pressure.[190,191] With a suitable BFH cycle, phenomena like vaporization, ion pair dissociation, solvation, and fusion can be connected, possibly opening the route to predictions of bulk physical properties. In an earlier study by our group, we could approximate the Gibbs enthalpy of fusion ($\Delta_{fus}G$) at the melting point as the sum of Gibbs lattice and solvation enthalpies (see eqn. (26)).[87]

(26) $\Delta_{fus}G(T_{fus}) = \Delta_{latt}G(T_{fus}) + \Delta_{solv}G(T_{fus})$

Here, $\Delta_{latt}G$ was calculated from standard VBT models[129,127] and the calculated (BP86/SV(P)) gas phase entropy; $\Delta_{solv}G$ was calculated with COSMO using the experimental dielectric constant.

For a more fundamental and precise treatment, we decided to assess all enthalpies in the full phase-transformation cycle of ionic liquids. It follows the equation:

(27) $\Delta_{latt}H = \Delta_{fus}H + \Delta_{vap}H + \Delta_{diss}H$

Given the knowledge of three quantities from this cycle, the fourth can be calculated. In view of the sparsely available data, we restricted ourselves to imidazolium ILs. Wherever possible, we used experimental data; if that was not the case, we made estimations as detailed below.

We created a test set of ILs for which as many experimental and calculated data as possible are available, making a compromise between high chemical variety and the need for approximations. The resulting set, for which the entire cycle will be closed, will be denoted as "set 0" (see Table 50). For the establishment of the enthalpies contributing to the above BFH cycle, however, other sets had to be used, as will be shown.

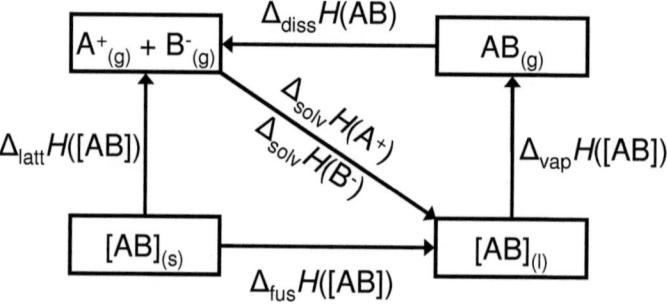

Figure 63: The extended Born-Fajans-Haber cycle for phase transformations of an ionic compound in terms of enthalpies. The cycle can also be closed using energies (ΔU), free energies (ΔG) or entropies (ΔS); however, most experimental data is available for enthalpies.

Table 50: The ion combinations of set 0.

cation	$[C_2MIm]^+$	$[C_3MIm]^+$	$[C_4MIm]^+$	$[C_6MIm]^+$	$[C_8MIm]^+$
anion	$[BF_4]^-$		$[BF_4]^-$		$[BF_4]^-$
	Cl^-	Cl^-	Cl^-		
	$[C_2SO_4]^-$				
	$[Dca]^-$	$[Dca]^-$	$[Dca]^-$		$[Dca]^-$
	$[NO_3]^-$	$[NO_3]^-$	$[NO_3]^-$		
	$[NTf_2]^-$		$[NTf_2]^-$	$[NTf_2]^-$	$[NTf_2]^-$
	$[OAc]^-$	$[OAc]^-$			
	$[PF_6]^-$	$[PF_6]^-$	$[PF_6]^-$		$[PF_6]^-$
	$[SCN]^-$	$[SCN]^-$	$[SCN]^-$		
	$[Tcm]^-$	$[Tcm]^-$	$[Tcm]^-$		

4.6.1. The Dissociation Enthalpy

The dissociation enthalpy ($\Delta_{diss}H$) is the energy needed to disrupt an ion pair in the gas phase into infinitely separated ions. Using mass spectroscopy, a qualitative scale can be constructed for dissociation enthalpies.[192] Calculated dissociation energies $\Delta_{diss}U$ are highly dependent on method and basis set. For weakly (hydrogen-)bound systems like the NH_3 dimer, computationally extremely expensive methods like Weizmann-1 and -2 are known to give excellent results.[193,194] For IL ion pairs, standard DFT functionals are lacking in their treatment of dispersive interactions;[195] however, dispersion-corrected functionals like BLYP-D[196,4,197] at the basis set limit, in combination with counterpoise correction, have proven to give very good accordance with the reference MP2 values.[32]

Table 51 collects newly calculated values alongside known ones. DFT and G3MP2 energies differ about 15...50 kJ mol^{-1} for the same compound, which is most likely caused by the lack of treatment of dispersive interactions by standard DFT functionals. As an analysis of Table 51 shows, even SCS-MP2[198] does not treat correlation well enough to fully describe dispersion in these systems.

Table 51: Calculated dissociation energies, $\Delta_{diss}U$ [kJ mol^{-1}], for some ILs.

IL	BP86/ TZVP	BP86/ TZVPP	B3LYP[42]/ TZVP	SCS-MP2[a]	G3MP2
[C$_2$MIm][Dca]	334	337	333	-/-	355
[C$_2$MIm][C$_2$SO$_4$]	355	358	355	-/-	387
[C$_2$MIm][NTf$_2$]	319[b]	313[b]	320[b]	333[b]	350
[C$_3$MIm][Dca]	333	336	332	-/-	351
[C$_4$MIm][Dca]	332	334	331	-/-	349
[C$_4$MIm][PF$_6$]	317	321	321	-/-	356
[C$_4$MIm][NTf$_2$]	315[b]	309[b]	317[b]	-/-	357

[a] SCS-MP2/ TZVPP // MP2/ TZVP. [b] Taken from ref. [183].

For the aforementioned reasons, all standard dissociation enthalpies used in this study were calculated with G3MP2.[199] The standard error σ for all G3MP2 energies of formation against experimental values (see Table III in ref. [44]) is 6.9 kJ mol^{-1} for the G3 set of molecules. Accordingly, there's a root-mean square error of 5.9 kJ mol^{-1} against the reference values (CCSD(T)/aug-cc-pVTZ[200,201]) for G3MP2 proton affinities of several IL anions.

Table 52: The extrapolation of dissociation enthalpies of short-chain 1-alkyl-3-methyl-imidazolium ILs (# = number of carbon atoms in the alkyl group) to longer chains, sorted by counteranion. σ$_{est}$ given in kJ mol^{-1}.

anion	extrapolated from (#)	extrapolated to (#)	r^2	σ$_{est}$
[BF$_4$]$^-$	2, 3, 4	8	0.8421	0.15
[Dca]$^-$	2, 3, 4	8	0.9749	0.69
[PF$_6$]$^-$	2, 3, 4	8	0.9382	0.20
[NTf$_2$]$^-$	2, 4	6, 8	1.0000	0.52[a]

[a] Expected error (see text).

Since dissociation enthalpies of imidazolium ILs with the same anion are heavily correlated with the length of the cation's alkyl chain, they were extrapolated to compounds with longer (> 4 C atoms) chains based on the values with shorter chains in order to reduce computational cost (see Table 52). As for [C₃MIm][Tcm], the mean value of [C₂MIm] and [C₄MIm][Tcm] dissociation enthalpies was taken.

In the cases where it applies, the errors introduced by this extrapolation need to be added to the G3MP2 error according to the formula for error propagation:

$$(28) \quad \sigma = \sqrt{\sum_i \sigma_i^2}$$

where σ is the total resulting error, σ_i are the individual errors of the methods; here, i = 2. For [NTf₂]⁻, we made an extrapolation from [C#MIm]⁺ with # = 2 and 4 to # = 6 and 8. For two data points, no σ_{est} can be calculated; therefore, we used the extrapolation error of the other anions according to the following formula to find the expected error:

$$(29) \quad \sigma = \sqrt{\sum_i \frac{\sigma_i^2}{N-1}}$$

where σ_i is the given error of the i-th determined value, N is the number of measurements. In this case, we expect an error of 0.52 kJ mol⁻¹ for the [NTf₂]⁻ extrapolations.

These additional errors, however, are insufficient to change the first two figures of σ_{est} (6.9 kJ mol⁻¹).

Table 53: Calculated (G3MP2) dissociation enthalpies in kJ mol⁻¹.

IL	$\Delta_{diss}H^0$	IL	$\Delta_{diss}H^0$
[C₂MIm][BF₄]	372	[C₃MIm][Tcm]	340
[C₂MIm][C₂SO₄]	385	[C₃MIm]Cl	395
[C₂MIm][Dca]	353	[C₄MIm][BF₄]	372
[C₂MIm][NO₃]	384	[C₄MIm][Dca]	347
[C₂MIm][NTf₂]	344	[C₄MIm][NO₃]	383
[C₂MIm][OAc]	426	[C₄MIm][NTf₂]	351
[C₂MIm][PF₆]	352	[C₄MIm][PF₆]	353
[C₂MIm][SCN]	371	[C₄MIm][SCN]	369
[C₂MIm][Tcm]	340	[C₄MIm][Tcm]	339
[C₂MIm]Cl	394	[C₄MIm]Cl	394
[C₃MIm][Dca]	349	[C₆MIm][NTf₂]	358[a]
[C₃MIm][NO₃]	383	[C₈MIm][BF₄]	371[a]
[C₃MIm][OAc]	426	[C₈MIm][Dca]	334[a]
[C₃MIm][PF₆]	352	[C₈MIm][NTf₂]	365[a]
[C₃MIm][SCN]	369	[C₈MIm][PF₆]	355[a]

[a] Extrapolated value (see text).

4.6.2. The Vaporization Enthalpy

The vaporization enthalpy is the energy needed for evaporating a liquid compound, destroying all near-ordering for uncharged molecules. For ILs, an ion pair is formed in the gas phase, which means the interaction energy of this pair is the only one preserved.[202] It can be determined using a multitude of experimental methods, or by MD simulations; overviews are given in ref. [203] and [204]. MD simulations also strongly suggest the formation of clusters on evaporation of ILs at room temperature.[203,205,206] However, this does not concern the BFH cycle as shown above, since at the elevated temperatures at which vaporization enthalpies are measured, single ion pairs prevail.

As for approximations and predictions, Kabo et al. found a relationship based upon the Stefan equation using the experimentally determined surface tension as well as the molecular volume.[207] A recent approach made use of calculated heats of formation (G3MP2) of the gas phase ion pair and the measured (by combustion calorimetry) heat of formation of the liquid phase.[30,208] Verevkin and Paulechka et al. found atomic contribution methods.[209,210] Jones et al. found a relationship based on the Coulomb and vdW interactions of cation and anion.[211] However, in this case, while the Coulomb interaction was calculated using the molecular radius and was based on *a priori* considerations, the vdW part was not, and was instead made to fit the remainder of the experimental vaporization enthalpy after subtracting the Coulomb part. Thus, its sizes are counterintuitive, as, for example, they do not correlate with the molecular surface at all as would be expected (see Table 54; r^2 = 0.0017).

Table 54: Anionic van-der-Waals contributions according to ref. [211] and calculated (with BP86/TZVP+COSMO) surfaces of different anions.

anion	$\hat{S}$ [nm^2]	$\Delta_{vdW}H^-$ [kJ mol^{-1}]
[SCN]$^-$	0.87	14
[BF$_4$]$^-$	0.91	13
[Dca]$^-$	1.02	18
[PF$_6$]$^-$	1.14	27
[OTf]$^-$	1.28	11
[C$_2$SO$_4$]$^-$	1.39	34
[FeCl$_4$]$^-$	1.52	44
[NTf$_2$]$^-$	2.06	13
[Fap]$^-$	2.59	18

In search for a fundamental expression, we tried to determine other factors that would play a role in the energetics of the vaporization process. It has been known for some time now that the vaporization enthalpy, $\Delta_{vap}H^0$, of classes of organic molecules depends on the solvent-accessible surface, $\hat{S}$.[212] The surface in the form of $V_m^{2/3}$ is also a constituent of $\Delta_{vap}H$ in the relationship given in ref. [207]. Thus, we set:

$$(30) \quad \Delta_{vap}H^0 = a \cdot V_m^{2/3} + b$$

For the chemically rather diverse ionic liquids we investigated, however, we found this simple correlation to be rather weak, with $r^2 = 0.1723$, $\sigma_{est} = 10.5$ kJ mol^{-1}. Upon including the standard gas phase enthalpy into the equation, i.e.

$$(31) \quad \Delta_{vap}H^0 = a \cdot V_m^{2/3} + b \cdot H_g^0 + c$$

we found a steep increase in r^2 to 0.8238, $\sigma_{est} = 4.8$ kJ mol^{-1}. Here, a = -224.4 kJ mol^{-1} nm^{-2}, b = 0.09288, c = 193.6 kJ mol^{-1}. As the expected experimental error is 4.7 kJ mol^{-1}, a further refinement by introducing additional descriptors is possibly inadequate.

All quantities were calculated as the sum of the single-ion values. H_g^0 is calculated by freeh with the following equation:

$$(32) \quad H_g = E_{ZP} + 4 \cdot RT + \sum_i \frac{v_i \left(1 + \exp\left(-\frac{v_i}{kT}\right)\right)}{2\left(1 - \exp\left(-\frac{v_i}{kT}\right)\right)}$$

where v_i is the i-th vibrational frequency.

For comparison, if the gas phase enthalpy is calculated for ion pairs, σ_{est} drops only by 9.6% which does not justify the very time-consuming optimization of the ion pair. The negative value of coefficient a might be explained by the energy gain in destroying a phase frontier (the liquid level). Enthalpy is present in the gas as well as in the liquid. However, its sizes do not exactly cancel out as additional degrees of freedom need to be excited on transfer to the gaseous state, which is denoted by the small positive value of coefficient b. An important part of the enthalpy is the zero-point energy which becomes important for rather light compounds with many degrees of vibrational freedom like the ones typically forming ILs.

Experimental and calculated values are given in Table 55. For comparison, we calculated $\Delta_{vap}H^0$ with COSMO-RS, using the optimized (BP86/TZVP) ion pair structures described in section 2.2.2 (designated $\Delta_{vap}H^0(C)$); here, $\sigma_{est} = 8.4$ kJ mol^{-1}, $r^2 = 0.4931$. Especially for the stronger coordinating nitrates, the error exceeds 15 kJ mol^{-1}. Figure 64 depicts these findings; the scatter for COSMO-RS values is quite high. Because of the tedious optimization of weakly coordinating ion pairs in the gas phase and the issues of COSMO-RS in this regard, we did not use this method any further.

Table 55: Calculated and experimental enthalpies of vaporization; all energies are given in kJ mol^{-1}, V_m is given in nm^3.

IL	V_m	H_g	calc. $\Delta_{vap}H^0$ (COSMO-RS)	calc. $\Delta_{vap}H^0$ (eqn. (31))	exp. $\Delta_{vap}H^0$	ref.
[C$_2$MIm][BF$_4$]	0.228	505	141	157	149	202
[C$_2$MIm][C$_2$SO$_4$]	0.282	687	161	161	164	202
[C$_2$MIm][NO$_3$]	0.211	501	146	160	168	29
[C$_2$MIm][NTf$_2$]	0.382	630	134	134	137	202, 205, 213
[C$_2$MIm][SCN]	0.224	486	151	156	151	211
[C$_4$MIm][Dca]	0.286	673	155	159	157	208
[C$_4$MIm][NO$_3$]	0.259	653	151	163	168	29
[C$_4$MIm][NTf$_2$]	0.430	782	135	138	136	202,205,213
[C$_4$MIm][Tcm]	0.318	704	167	155	156	209
[C$_6$MIm][NTf$_2$]	0.477	934	142	143	141	202,205,213
[C$_8$MIm][BF$_4$]	0.371	962	156	167	162	202
[C$_8$MIm][Dca]	0.381	978	170	167	162	211
[C$_8$MIm][NTf$_2$]	0.525	1087	151	148	151	202,205,213
[C$_8$MIm][PF$_6$]	0.402	979	161	162	169	202

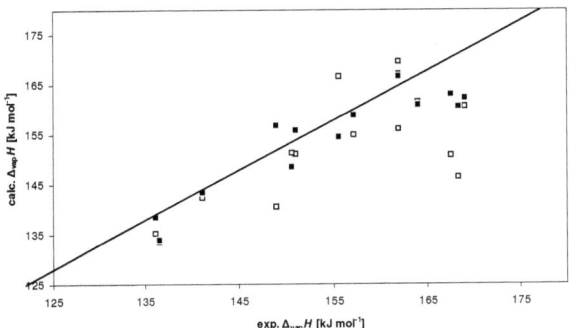

Figure 64: The comparison of experimental and calculated vapor pressures. Filled squares belong to data from eqn. (31), hollow ones to data from COSMO-RS.

For the proof of eqn. (31), we calculated further enthalpies of vaporization and compared them with their experimental counterparts. Results are given in Table 56. The errors are small in most cases (see below); for [C$_3$C$_1$MIm][NTf$_2$], [C$_6$Py][NTf$_2$], and [C$_8$MIm][OTf], they are relatively high ($\approx$10 kJ mol^{-1}), the latter probably due to the strong coordination of cation and anion. For the values for which the average of measurements given in multiple references was assumed, the average individual error σ was calculated with eqn. (29). For all ILs given in Table 55, the expected experimental error then is 2.9 kJ mol^{-1}. Added to the calculative error of 4.8 kJ mol^{-1} via error propagation (eqn. (28)), a total error of 5.6 kJ mol^{-1} for the interpolation given in eqn. (31) results.

Table 56: Further calculated (according to eqn. (31)) and experimental enthalpies of vaporization; all energies are given kJ mol^{-1}, V_m is given in nm^3. The [C$_4$MIm][FeCl$_4$] outlier is given in italics.

IL	V_m	H_g	calc. $\Delta_{vap}H^0$ (eqn. (31))	exp. $\Delta_{vap}H^0$	ref.
[C$_2$MIm][NPf$_2$]	0.461	707	125	136	213
[C$_3$C$_1$MIm][NPf$_2$]	0.509	858	130	142	213
[C$_3$C$_1$MIm][NTf$_2$]	0.430	781	138	151	213
[C$_4$MIm][FeCl$_4$]	*0.353*	*641*	*141*	*170*	*211*
[C$_4$MIm][NPf$_2$]	0.509	860	130	135	213
[C$_6$MIm][NPf$_2$]	0.555	1011	136	139	213
[C$_8$MIm][NPf$_2$]	0.603	1164	141	145	213
[C$_8$MIm][OTf]	0.418	1001	161	151	202
[C$_{10}$MIm][NTf$_2$]	0.572	1239	154	155	213
[C$_{10}$MIm][NPf$_2$]	0.650	1316	147	148	213
[C$_4$MPyr][Dca]	0.301	827	170	160	211
[C$_4$MPyr][Fap]	0.547	1057	142	152	211
[C$_4$MPyr][NTf$_2$]	0.446	935	150	152	211
[C$_6$MPyr][NTf$_2$]	0.503	1088	153	156	211
[C$_8$MPyr][NTf$_2$]	0.548	1240	159	161	211
[C$_4$Py][BF$_4$]	0.270	629	158	167	211
[C$_6$Py][NTf$_2$]	0.472	905	142	152	211

It should be noted that, although eqn. (31) is calibrated for [C$_\#$MIm]-type ILs only, it works surprisingly well for also for pyrrolidinium and pyridinium ILs. The paramagnetic [C$_4$MIm][FeCl$_4$] poses an exception with an error of about 30 kJ mol^{-1}, likely due to problems with the DFT and COSMO-RS calculations.[214] Nevertheless, we think that for the interpolation of missing enthalpies of vaporization of diamagnetic compounds akin to those in set 0, eqn. (31) will yield acceptable values.

4.6.3. The Fusion Enthalpy

The enthalpy of fusion ($\Delta_{fus}H$) is the energy needed to turn a crystal lattice to an isotropic liquid phase, i.e. the destruction of far-ordering. It can be determined experimentally using standard calorimetric methods, or calculated (for typical uncharged organic molecules) using a group contribution method.[215,216]

Fusion enthalpies are not as dependent on impurities as melting points.[160] Table 57 shows all imidazolium ILs for which a value was available as of this writing. Few trends are observable; [NTf$_2$]$^-$ compounds generally tend to have higher fusion enthalpies, while sulfates and fluorometallates are rather at the lower end of the spectrum.

The range of values is very small; the average is 21.4 kJ mol^{-1}, with a standard deviation of only 5.2 kJ mol^{-1}. The underlying trends in a narrow range of small values, like $\Delta_{fus}H$, usually get entirely drowned out by the errors made in the calculation of their constituting energies; also, the experimental errors become rather large in comparison to both the total values and their range. Thus, there seems to be little sense in trying to find a more general formula. For a manifold of linear combinations of physicochemical descriptors, no significant correlation could be found ($r^2 \leq 0.03$). Thus, we needed to fall back on using the average value. The expected experimental error for all measurements in Table 57 is 1.1 kJ mol^{-1}; added up via eqn. (28), the total error in assuming the average value is 5.4 kJ mol^{-1}.

To normalize the enthalpies, which are always measured at the experimental melting point, without knowing the heat capacity in the liquid and solid state, there's Sidgwick's equation:[217]

$$(33) \quad \Delta_{fus}H^0 = \Delta_{fus}H(T_{fus}) + (298 \text{ K} - T_{fus}) \cdot 54.4 \text{ J mol}^{-1} \text{ K}^{-1}$$

Table 57: Different imidazolium ILs and their experimental enthalpies of fusion (in kJ mol^{-1}).

IL	$\Delta_{fus}H$	ref.	IL	$\Delta_{fus}H$	ref.
[C$_2$MIm][BF$_4$]	9.5±1.2	218	[C$_5$MIm][NTf$_2$]	22.5	219
[C$_4$MIm][C$_8$SO$_4$]	12.7±2.6	220	[C$_4$MIm][NTf$_2$]	22.9±1.0	160,31,221
[C$_1$MIm][C$_1$SO$_4$]	16.6±0.5	222	[C$_4$MIm]Br	22.9±0.1	173
[C$_2$MIm][PF$_6$]	17.9±0.7	223,224	[C$_1$MIm][NTf$_2$]	24.5±0.8	221
[C$_4$MIm][NO$_3$]	18.0±0.2	186	[C$_2$MIm][NTf$_2$]	22.8±1.7	219,221,185
[C$_2$MIm]Br	18.3±0.2	173	[C$_8$MIm][NTf$_2$]	25.2±0.3	185
[C$_4$MIm][OTfa]	19.1±0.1	171	[C$_2$MIm][NO$_3$]	26.7	this study
[C$_3$C$_1$MIm][NTf$_2$]	19.7±2.5	218	[C$_6$MIm][NTf$_2$]	28.3±0.2	188
[C$_4$MIm][PF$_6$]	19.8±0.6	160,163	[C$_2$MIm][OTos]	27.8±0.8	222,225
[C$_4$MIm][OTos]	21.6±0.2	187	[C$_8$MIm][Dca]	30.7	this study

If applied to all data from set 0 where melting point as well as melting enthalpy are given (see Table 58), we get an average correction to $\Delta_{fus}H^0$ of 0.4±1.4 kJ mol^{-1}. However, this correction is of the same order of magnitude as the experimental uncertainty; to ignore it, i.e. to set $\Delta_{fus}H^0 = \Delta_{fus}H(T_{fus})$, may result in better values,[226] which is what we will be doing. However, we need to be aware that in this way, we will introduce an – though rather small – additional error.

Table 58: Melting points and enthalpies of fusion. To ensure best quality, melting points were taken from the batch with lowest water content, wherever possible.

IL	T_{fus} [K] (exp.)	$\Delta_{fus}H$ [kJ mol^{-1}] (exp.)	$\Delta_{fus}H^0$ [kJ mol^{-1}] (eqn. (33))
[C$_2$MIm][BF$_4$]	286.2[227]	9.5	10.2
[C$_2$MIm][PF$_6$]	333.3[223,224,228]	17.9	16.0
[C$_2$MIm][NO$_3$]	320.1^a	26.7	25.5
[C$_2$MIm][NTf$_2$]	254.2[87]	22.8	25.1
[C$_4$MIm][NO$_3$]	309.2[186]	18.0	17.4
[C$_4$MIm][NTf$_2$]	268.2[87]	22.9	24.5
[C$_4$MIm][PF$_6$]	282.2[87]	19.8	20.6
[C$_6$MIm][NTf$_2$]	267.2[229]	28.3	28.4
[C$_8$MIm][Dca]	287.0^a	30.7	31.3
[C$_8$MIm][NTf$_2$]	264.0[185]	25.2	27.0

^{a} Measured for this study.

4.6.4. The Solvation Enthalpy

The solvation enthalpy ($\Delta_{solv}H$), as we use it in this study, is the change in enthalpy upon forming a liquid phase from two indefinitely separated ions in the gas phase. It cannot be directly measured, but inferred from the difference of fusion and lattice enthalpy. However, it may be calculated using one out of many solvent models (e.g. COSMO, IPCM and SCI-PCM[153]) which are mostly dependent on the dielectric constant ε_r of the liquid as input.[88] We investigated two possibilities to calculate $\Delta_{solv}H^0$:

(a) By using the BFH cycle shown in Figure 63:

(34) $\Delta_{solv}H^0 = -(\Delta_{vap}H^0 + \Delta_{diss}H^0)$

where $\Delta_{vap}H^0$ was taken from experimental data where available, eqn. (30) otherwise, and $\Delta_{diss}H^0$ from G3MP2 calculations. To our knowledge, this is the first account on an assessment of solvation enthalpies of IL ions on the basis of experimental data where possible. These numbers are considered to be the most reliable currently available and should be the benchmark for comparing them to other methodologies.

(b) With COSMO-RS, where the IL is defined as a 1:1 mixture of anion and cation. The single ion solvation enthalpy of the individual ions is part of COSMOtherm's output, albeit there given as the negative vaporization enthalpy.

Table 59: Calculated enthalpies of solvation (given in kJ mol^{-1}).

IL	$\Delta_{solv}H^0$ (eqn. (34))	$\Delta_{solv}H^0$ (COSMO-RS)	IL	$\Delta_{solv}H^0$ (eqn. (34))	$\Delta_{solv}H^0$ (COSMO-RS)
[C$_2$MIm][BF$_4$]	-521±7.2	-493	[C$_3$MIm][Tcm]	-462±8.9	-467
[C$_2$MIm][C$_2$SO$_4$]	-549±8.0	-528	[C$_3$MIm]Cl	-557±8.9	-579
[C$_2$MIm][Dca]	-506±8.9	-496	[C$_4$MIm][BF$_4$]	-532±8.9	-493
[C$_2$MIm][NO$_3$]	-552±7.5	-527	[C$_4$MIm][Dca]	-504±7.0	-496
[C$_2$MIm][NTf$_2$]	-481±7.2	-456	[C$_4$MIm][NO$_3$]	-550±7.5	-529
[C$_2$MIm][OAc]	-591±8.9	-559	[C$_4$MIm][NTf$_2$]	-487±7.3	-455
[C$_2$MIm][PF$_6$]	-503±8.9	-469	[C$_4$MIm][PF$_6$]	-507±8.9	-467
[C$_2$MIm][SCN]	-522±7.2	-523	[C$_4$MIm][SCN]	-527±8.9	-524
[C$_2$MIm][Tcm]	-491±8.9	-468	[C$_4$MIm][Tcm]	-495±7.5	-468
[C$_2$MIm]Cl	-556±8.9	-577	[C$_4$MIm]Cl	-558±8.9	-582
[C$_3$MIm][Dca]	-506±8.9	-495	[C$_6$MIm][NTf$_2$]	-499±7.1	-460
[C$_3$MIm][NO$_3$]	-545±8.9	-527	[C$_8$MIm][BF$_4$]	-533±7.5	-504
[C$_3$MIm][OAc]	-592±8.9	-560	[C$_8$MIm][Dca]	-496±8.0	-508
[C$_3$MIm][PF$_6$]	-505±8.9	-466	[C$_8$MIm][NTf$_2$]	-515±7.1	-466
[C$_3$MIm][SCN]	-526±8.9	-522	[C$_8$MIm][PF$_6$]	-524±8.0	-476

We found the expected error for method (a) to be 8.4 kJ mol^{-1} (see Table 59). With respect to method (a), the values calculated with method (b) have a σ of 28.2 kJ mol^{-1}. This means that eqn. (34) seems much more valuable than COSMO-RS for the calculation of solvation enthalpies, but needs the tedious computation of the dissociation enthalpy.

As mentioned above, a third method would be the employment of a classical solvation model using the experimental dielectric constant. For those compounds where a dissociation enthalpy as well as ε_r are available, we calculated the Gibbs solvation energy $\Delta_{solv}G$ with COSMO and converted it with the following formula to obtain standard conditions:[230]

(35) $\Delta_{solv}G^0 = \Delta_{solv}G + RT \ln(\rho/M/c_0) - 7.93$ kJ mol^{-1}

where $c_0 = 1$ mol l^{-1}; M is the molecular mass; the density ρ was calculated with equations (10), (12) and (16). The enthalpy of solvation can then be calculated with:

(36) $\Delta_{solv}H^0 = \Delta_{solv}G^0 - T\Delta S$

where $\Delta S = S_l^0 - S_g^0$, $T = 298.15$ K. The liquid standard entropy was calculated using eqn. (21), the gas phase standard entropy from a BP86/SV(P) geometry optimization of the single ions and vibrational analysis with AOFORCE.

Table 60: Calculated solvation enthalpies according to eqn. (34) and (38), and the constituents needed for their calculation. The molecular mass is given in g mol^{-1}, all other quantities in kJ mol^{-1}; T = 298.15 K.

IL	M	$T\Delta S$	ε_r	$\Delta_{solv}H^0$ (eqn. (34))	$\Delta_{solv}H^0$ (eqn. (36))
[C$_2$MIm][BF$_4$]	198	85.4	12.8	-521	-503
[C$_2$MIm][NTf$_2$]	391	88.8	12.3	-481	-456
[C$_4$MIm][NTf$_2$]	419	86.7	11.6	-487	-444
[C$_4$MIm][BF$_4$]	226	87.8	11.7	-532	-496
[C$_4$MIm][PF$_6$]	284	76.5	11.4	-507	-470

Results are given in Table 60. Compared to the BFH values, COSMO yielded σ = 37.1 kJ mol^{-1}, r^2 = 0.8455; for the same set of compounds, the correlation of eqn. (34) with the COSMO-RS values (see Table 59) would give σ = 37.2 kJ mol^{-1}, r^2 = 0.9111, meaning that COSMO-RS and COSMO solvation enthalpies have about the same quality; COSMO-RS, however, has no need for an experimental dielectric constant.

We found that there seems to be a connection between the calculated HOMO-LUMO gap (BP86/TZVP) of the ion pairs and the solvation enthalpy. Figure 65 shows this connection for the investigated ILs.

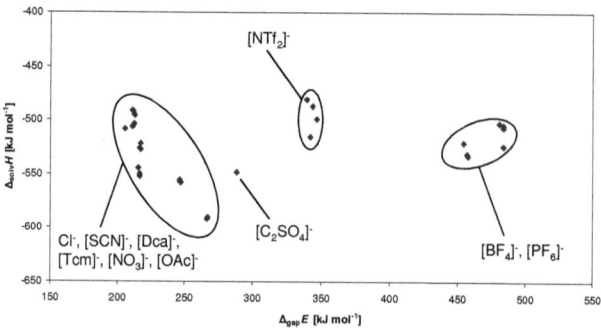

Figure 65: The correlation of $\Delta_{solv}H^0$ and $\Delta_{gap}E$ for IL ion pairs with different anions.

In our test set, all compounds with anions Cl⁻, [Dca]⁻, [OAc]⁻, [SCN]⁻ and [Tcm]⁻ form a series (see Figure 66) and obey the equation:

(37) $\Delta_{solv}H^0 = -1.56 \cdot \Delta_{gap}E - 175$ kJ mol⁻¹

Here, $\Delta_{solv}H^0$ was calculated using eqn. (34), while $\Delta_{gap}E$ was taken from a gas phase BP86/TZVP ion pair calculation. Then, r^2 = 0.9299, σ_{est} = 5.9 kJ mol⁻¹. This is a better correlation than the COSMO-RS values give, but is restricted to this type of anions. Naturally, the calculated HOMO-LUMO gap is dependent on quantum chemical method and basis set, accounting for some of the scatter. These anions are symmetric, chemically rather soft, and, except for chloride, share pronounced π-delocalization over the entire molecule, leaving them with little possibility for dipole-dipole interaction in the ion pair. In contrast, [BF₄]⁻ and [PF₆]⁻ show σ-delocalization, [NTf₂]⁻ shows both, and the delocalized charge of [C₂SO₄]⁻ is counterbalanced by localized ethyl electrons. Also, the latter two own distinct dipoles; additionally, [BF₄]⁻, [PF₆]⁻ and [NTf₂]⁻ are little polarizable. Nitrate ILs are probably stronger coordinating and therefore present slight outliers. We assume σ-σ* type bonding between the anion and the acidic hydrogen of the cation. Since the cationic part of this bond stays more or less the same, its strength is proportional to the amount of interaction with the anionic dipole. Because we are not investigating ion pairs but the transition from ideally separated ions in the gas phase to an isotropic liquid, the backbonding (i.e. charge transfer) measured by $\Delta_{gap}E$ of the paired ions represents σ-σ* as well as π-σ* and π-π* bonding ability between statistically coordinated anions and cations, all of which should be present in a rather dense, disordered liquid with π-delocalized ions.

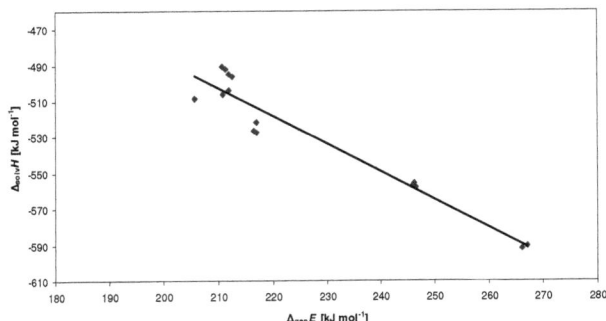

Figure 66: The connection of $\Delta_{solv}H^0$ and $\Delta_{gap}E$ for certain ILs (see text).

4.6.5. The Lattice Enthalpy

For ionic compounds, the lattice enthalpy is the energy needed for the transfer from its densest state, i.e. the crystal lattice, to infinitely separated ions in the gas phase, resulting in the total destruction of any near- or far-ordering. Therefore, it may be seen as a measure of the total stability of a lattice. Except for using a BFH cycle as given in Figure 63 and adding up the terms according to eqn. (27), the lattice enthalpy may also be determined experimentally using enthalpies of formation, or simulated using a crystal structure as starting point. The article in ref. [129] gives an overview of the latter two possibilities. Another possibility lies in using an appropriate VBT formula:[130]

(38) $\Delta_{latt}U = a \cdot V_m^{-1/3} + b$

with a = 234.6 kJ mol^{-1} nm^{-1}, b = 103.8 kJ mol^{-1};[129] and

(39) $\Delta_{latt}H = \Delta_{latt}U + \left(\dfrac{n^+ + n^- - 8}{2}\right) \cdot RT$

where n$^\pm$ = 3 for monoatomic, 5 for linear and 6 for nonlinear ions.[231]

Table 61: Enthalpies of the large BFH cycle (in kJ mol^{-1}; three significant figures) alongside the molecular volume, sorted by cation and $\Delta\Delta_{latt}H$ (the difference of columns 6 and 7). Experimental $\Delta_{vap}H^0$ were taken from the given references. G3MP2-Dissociation enthalpies from Table 53 are included for comparison.

IL	V_m [nm³]	$\Delta_{diss}H^0$	$\Delta_{vap}H^0$	$\Delta_{fus}H^{(0)}$	$\Delta_{latt}H^0$ (eqn. (27))	$\Delta_{latt}H^0$ (eqn. (39))	$\Delta\Delta_{latt}H^0$
[C₂MIm][BF₄]	0.228	372±6.9	149±2.0[202]	9.54±1.2	531±7.3	493	38.1
[C₂MIm][C₂SO₄]	0.282	385±6.9	164±4.0[202]	21.4±5.4[b]	570±9.6	467	104
[C₂MIm][Dca]	0.238	353±6.9	156±5.6[a]	21.4±5.4[b]	530±10.4	487	42.6
[C₂MIm][NO₃]	0.211	384±6.9	168±2.9[29]	26.7±1.1	579±7.6	503	76.3
[C₂MIm][NTf₂]	0.382	344±6.9	137±2.2 [202,205,213]	22.8±1.7	504±7.4	432	71.6
[C₂MIm][OAc]	0.227	426±6.9	165±5.6[a]	21.4±5.4[b]	612±10.4	493	119
[C₂MIm][PF₆]	0.260	352±6.9	151±5.6[a]	17.9±0.7	521±8.9	476	44.3
[C₂MIm][SCN]	0.224	371±6.9	151±2[211]	21.4±5.4[b]	543±9.0	494	49.6
[C₂MIm][Tcm]	0.270	340±6.9	151±5.6[a]	21.4±5.4[b]	512±10.4	469	43.0
[C₂MIm]Cl	0.190	394±6.9	162±5.6[a]	21.4±5.4[b]	577±10.4	514	63.0
[C₃MIm][Dca]	0.262	349±6.9	157±5.6[a]	21.4±5.4[b]	527±10.4	475	52.0
[C₃MIm][NO₃]	0.235	383±6.9	162±5.6[a]	21.4±5.4[b]	566±10.4	489	77.2
[C₃MIm][OAc]	0.251	426±6.9	166±5.6[a]	21.4±5.4[b]	613±10.4	480	133
[C₃MIm][PF₆]	0.284	352±6.9	152±5.6[a]	21.4±5.4[b]	526±10.4	466	60.2
[C₃MIm][SCN]	0.248	369±6.9	157±5.6[a]	21.4±5.4[b]	548±10.4	481	67.0
[C₃MIm][Tcm]	0.294	340±6.9	153±5.6[a]	21.4±5.4[b]	514±10.4	462	52.0
[C₃MIm]Cl	0.214	395±6.9	163±5.6[a]	24.9±5.4[b]	579±10.4	498	80.3
[C₄MIm][BF₄]	0.276	372±6.9	160±5.6[a]	21.4±5.4[b]	553±10.4	469	83.7
[C₄MIm][Dca]	0.286	347±6.9	157±1.1[208]	21.4±5.4[b]	525±8.8	465	60.3
[C₄MIm][NO₃]	0.259	383±6.9	168±2.9[29]	18.0±0.2	568±7.5	477	91.4
[C₄MIm][NTf₂]	0.430	351±6.9	136±2.5 [202,205,213]	22.9±1.0	510±7.4	420	90.6
[C₄MIm][PF₆]	0.307	353±6.9	154±5.6[a]	19.8±0.6	527±8.9	456	70.6

[C$_4$Mim][SCN]	0.272	369±6.9	159±5.6[a]	21.4±5.4[b]	549±10.4	470	79.0
[C$_4$Mim][Tcm]	0.318	339±6.9	156±2.9[209]	21.4±5.4[b]	516±9.2	453	63.7
[C$_4$Mim]Cl	0.238	394±6.9	164±5.6[a]	21.4±5.4[b]	580±10.4	485	94.6
[C$_6$Mim][NTf$_2$]	0.477	358±6.9	141±1.6 [202,205,213]	28.1±0.2	527±7.1	409	118
[C$_8$Mim][BF$_4$]	0.371	371±6.9	162±3.0[202]	21.4±5.4[b]	554±9.2	435	119
[C$_8$Mim][Dca]	0.381	334±6.9	162±4.0[211]	30.7±1.1	527±8.1	432	94.4
[C$_8$Mim][NTf$_2$]	0.525	365±6.9	151±1.6 [202,205,213]	25.2±0.3	541±7.1	400	141
[C$_8$Mim][PF$_6$]	0.402	355±6.9	169±4.0[202]	21.4±5.4[b]	546±9.6	427	119

[a] Approximated value (see eqn. (31)). [b] Average value; all other $\Delta_{fus}H$ values were taken from Table 57.

Table 61 shows the energies of the large BFH cycle, together with both values for the lattice enthalpy. The BFH lattice enthalpies established with our "best values" are about 40...140 kJ mol^{-1} higher than the VBT enthalpies, which is much higher than the combined errors of measurements and calculations (9.5 kJ mol^{-1} expected).

From the analysis of Table 61 we conclude that the usual VBT-type lattice energy formula (eqn. (38)) is not suitable for ILs. We therefore made a reassessment according to the following scheme: Using the BFH cycle, we calculated $\Delta_{latt}H^0$ and then employed eqn. (39) to obtain $\Delta_{latt}U$. With this, we started a general reparametrization of VBT-type equations according to Kapustinskii, Bartlett and others, i.e.:[232,129,233,130]

(40) $\Delta_{latt}U = a \cdot r_m^{-1} + b \cdot r_m^{-2} + c$

and eqn. (38), respectively. However, the errors σ_{est} obtained from these formulas (with optimized coefficients) are 22.7 and 25.2 kJ mol^{-1} still, which is not much below the standard deviation of $\Delta_{latt}U$ (29.4 kJ mol^{-1}), calculated with eqn. (27).

This implies that the current VBT formulas greatly underestimate the lattice energy or enthalpy. Since they were deduced for hardly polarizable, classical salts where charges are strongly localized and rather spherical (e.g. NaCl), they are lacking any kind of description of the dispersive interaction for the covalent molecular complex ions constituting the typical IL crystals. Therefore, a more systematic approach had to be considered. We investigated the lattice enthalpy and not the inner energy; both are easily convertible (see eqn. (39)). It can be expected that the lattice enthalpy would be heavily correlated with the solvation enthalpy, since both differ only by the fusion enthalpy, which is in a fairly narrow range for ionic liquids (see above). Because the dielectric constant for most ILs is unknown, we calculated the Gibbs solvation energy in an electrical conductor ($\Delta_{solv}G^\infty$) with COSMO ($\varepsilon_r = \infty$) and corrected it to a standard enthalpy ($\Delta_{solv}H^{0,\infty}$) according to eqn. (35) and (36). So, in a first move, we set:

(41) $\Delta_{latt}H^0 = a \cdot \Delta_{solv}H^{0,\infty} + b$

Since COSMO is a polarizable continuum model for solvation, we expected that a certain amount of dispersive interaction with the environment would already be included intrinsically. In fact, the correlation is a great deal better than traditional VBT, with r^2 = 0.8279, σ_{est} = 12.0 kJ mol^{-1}; a = -0.658, b = 171 kJ mol^{-1}.

Here, it seems that electrostatic interactions, which are adequately described by a reciprocal distance term in eqn. (38) and (40) for the more or less point-shaped charges of classical salts, are much better described with $\Delta_{solv}H^{0,\infty}$ for the irregularly shaped charges of organic salts. Indeed, the solvation enthalpy in an electrical conductor can be viewed as a measure for polarity in terms of local charge density of the individual ions, indicating dipole-charge and dipole-dipole interactions. Polarity, however, also includes dispersive interactions, meaning – consistent with the framework of the BFH cycle – that the largest part of the lattice enthalpy is included with the solvation enthalpy.

There might be a manifold of – more or less weak – interactions that comprise the far-ordering of the lattice; generally, they will grow in three dimensions of a molecule and thus can be approximated with the cubed molecular radius r_m^3. A further contribution would arise from the destruction of the crystal surface (which would yield energy) and scales with the solvent-accessible surface $\hat{S}$. As we move from the lattice to the gas phase, there will also be a direct contribution stemming from the standard gas phase enthalpy H_g^0. Thus, the resulting expression is:

(42) $\Delta_{latt}H^0 = a \cdot \Delta_{solv}H^{0,\infty} + b \cdot H_g^0 + c \cdot r_m^3 + d \cdot \hat{S} + e$

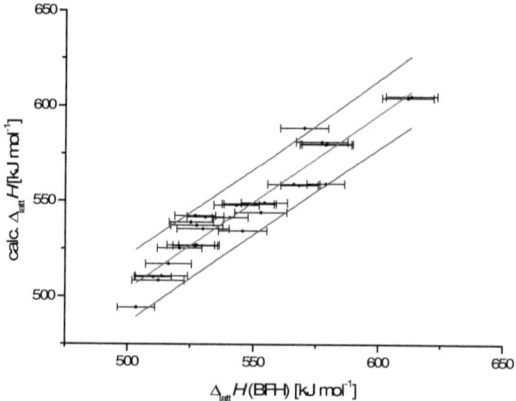

Figure 67: The correlation of lattice enthalpies taken from eqn. (27) (the BFH cycle) and (42), shown along with error bars and the 95% prediction bands.

Table 62: The calculated lattice enthalpies, according to eqn. (27) and (42), and its constituents (also needed for eqn. (35) and (36)). The molecular mass is given in g mol^{-1}, r_m in nm, $\hat{S}$ in nm^2, and all other quantities in kJ mol^{-1}; T = 298.15 K.

IL	$\Delta_{solv}G^{\infty}$	M	$T\Delta S$	H_g^0	r_m	$\hat{S}$	calc. $\Delta_{latt}H^0$ (eqn. (27))	calc. $\Delta_{latt}H^0$ (eqn. (42))
[C$_2$MIm][BF$_4$]	-475	198	85.4	505	0.595	2.53	531±7.3	542
[C$_2$MIm][C$_2$SO$_4$]	-513	236	102	687	0.646	3.01	570±9.6	589
[C$_2$MIm][Dca]	-466	177	91.6	521	0.606	2.64	530±10.4	535
[C$_2$MIm][NO$_3$]	-502	173	85.2	501	0.575	2.38	579±7.6	559
[C$_2$MIm][NTf$_2$]	-431	391	88.8	630	0.711	3.68	504±7.4	494
[C$_2$MIm][OAc]	-546	170	94.5	589	0.594	2.52	612±10.4	604
[C$_2$MIm][PF$_6$]	-449	256	84.8	522	0.627	2.77	521±8.9	525
[C$_2$MIm][SCN]	-484	169	93.8	486	0.591	2.49	543±9.0	548
[C$_2$MIm][Tcm]	-436	201	89.3	552	0.636	2.94	512±10.4	509
[C$_2$MIm]Cl	-550	147	79.6	456	0.543	2.15	577±10.4	581
[C$_3$MIm][Dca]	-464	191	92.1	597	0.623	2.84	527±10.4	537
[C$_3$MIm][NO$_3$]	-500	187	86.3	577	0.591	2.58	566±10.4	559
[C$_3$MIm][OAc]	-543	184	95.3	665	0.611	2.73	613±10.4	605
[C$_3$MIm][PF$_6$]	-447	270	82.9	598	0.643	2.97	526±10.4	527
[C$_3$MIm][SCN]	-481	183	94.4	562	0.608	2.69	548±10.4	549
[C$_3$MIm][Tcm]	-434	215	87.2	628	0.652	3.14	514±10.4	511
[C$_3$MIm]Cl	-548	161	82.5	532	0.560	2.35	579±10.4	580
[C$_4$MIm][BF$_4$]	-473	226	87.8	657	0.627	2.93	553±10.4	544
[C$_4$MIm][Dca]	-463	205	93.3	673	0.638	3.04	525±8.8	539
[C$_4$MIm][NO$_3$]	-500	201	86.4	653	0.607	2.79	568±7.5	559
[C$_4$MIm][NTf$_2$]	-429	419	86.7	782	0.743	4.08	510±7.4	510
[C$_4$MIm][PF$_6$]	-446	284	76.5	674	0.659	3.17	527±8.9	527
[C$_4$MIm][SCN]	-481	197	93.4	639	0.623	2.89	549±10.4	549
[C$_4$MIm][Tcm]	-434	229	94.0	704	0.668	3.34	516±9.2	517
[C$_4$MIm]Cl	-548	175	88.0	608	0.575	2.55	580±10.4	580
[C$_6$MIm][NTf$_2$]	-429	447	91.3	934	0.769	4.47	527±7.1	527
[C$_8$MIm][BF$_4$]	-473	282	99.8	962	0.677	3.72	554±9.2	549
[C$_8$MIm][Dca]	-464	261	97.3	978	0.688	3.84	527±8.1	542
[C$_8$MIm][NTf$_2$]	-430	475	96.4	1087	0.793	4.87	541±7.1	541
[C$_8$MIm][PF$_6$]	-447	340	80.9	979	0.709	3.96	546±9.6	534

Inclusion of these factors caused the error to drop to 8.4 kJ mol^{-1} (r^2 = 0.9153), which is probably too close to the experimental error to be further improved upon. All quantities were calculated separately for cation and anion and then added up. For H_g^0, see section 4.6.2. Here, a = -0.500, b = 0.272, c = 19.3 nm^{-3} MJ mol^{-1}, d = -196 nm^{-2} kJ mol^{-1}, e = 386 kJ mol^{-1}. Results are given in Table 62 and Figure 67.

We also found the solvation enthalpy calculated using the BFH cycle (eqn. (34)) to be correlated with $\Delta_{solv}H^{0,\infty}$ with $r^2 = 0.8420$ according to:

(43) $\Delta_{solv}H^0 = 0.700 \cdot \Delta_{solv}H^{0,\infty} - 128$ kJ mol^{-1}

This means we can almost fully substitute $\Delta_{solv}H^{0,\infty}$ in eqn. (42) for $\Delta_{solv}H^0$ and it becomes clear that the lattice enthalpy consists of the solvation enthalpy and an additional contribution ($b \cdot H_g + c \cdot r_m^3 + d \cdot \hat{S}$), which, according to Figure 63, would be needed to describe the fusion enthalpy. A small residue of $\Delta_{solv}H^{0,\infty}$ would still remain (since the coefficients 0.500 and 0.700 don't cancel out completely) and possibly forms a part of the enthalpy of fusion where it may stand for electrostatic interactions which are weakened at the destruction of the lattice. However, as stated in section 4.6.3, the experimental range of fusion enthalpies is too small to make any meaningful prediction of $\Delta_{fus}H$.

Using the best formulas for lattice and vaporization enthalpies (eqn. (31) and (42)), the dissociation enthalpy could be calculated according to the BFH cycle in Figure 63. Indeed, the equation is identical to eqn. (42), but with parameters a = -0.483, b = 0.164, c = 18.6 nm^{-3} MJ mol^{-1}, d = -164 nm^{-2} kJ mol^{-1}, e = 202 kJ mol^{-1}.

(44) $\Delta_{diss}H^0 = a \cdot \Delta_{solv}H^{0,\infty} + b \cdot H_g^0 + c \cdot r_m^3 + d \cdot \hat{S} + e$

As Table 63 shows, the chain length dependency of the dissociation enthalpy, as stated in section 4.6.1, is being correctly treated. Results are depicted in Figure 68. The error σ_{est} with respect to the G3MP2 values amounts to 8.0 kJ mol^{-1}; $r^2 = 0.8845$. The combined error of G3MP2 and eqn. (44) would then be 10.9 kJ mol^{-1} with respect to the experimental values, which is only 1.58 times the error inherent to G3MP2. With this method, however, no ion pairs need to be calculated. This is a considerable advantage, since all underlying calculations could be done on a time scale of hours rather than (several) months on common equipment.

Table 63: Calculated dissociation enthalpies (according to G3MP2 and eqn. (44), respectively) in kJ mol^{-1}.

IL	calc. $\Delta_{diss}H^0$ (G3MP2)	calc. $\Delta_{diss}H^0$ (eqn. (44))	IL	calc. $\Delta_{diss}H^0$ (G3MP2)	calc. $\Delta_{diss}H^0$ (eqn. (44))
[C$_2$MIm][BF$_4$]	372	366	[C$_3$MIm][Tcm]	340	339
[C$_2$MIm][C$_2$SO$_4$]	385	405	[C$_3$MIm]Cl	395	396
[C$_2$MIm][Dca]	353	361	[C$_4$MIm][BF$_4$]	372	363
[C$_2$MIm][NO$_3$]	384	380	[C$_4$MIm][Dca]	347	359
[C$_2$MIm][NTf$_2$]	344	336	[C$_4$MIm][NO$_3$]	383	374
[C$_2$MIm][OAc]	426	418	[C$_4$MIm][NTf$_2$]	351	347

IL	calc. $\Delta_{diss}H^0$ (G3MP2)	calc. $\Delta_{diss}H^0$ (eqn. (44))	IL	calc. $\Delta_{diss}H^0$ (G3MP2)	calc. $\Delta_{diss}H^0$ (eqn. (44))
[C$_2$MIm][PF$_6$]	352	354	[C$_4$MIm][PF$_6$]	353	351
[C$_2$MIm][SCN]	371	373	[C$_4$MIm][SCN]	369	368
[C$_2$MIm][Tcm]	340	339	[C$_4$MIm][Tcm]	339	343
[C$_2$MIm]Cl	394	399	[C$_4$MIm]Cl	394	393
[C$_3$MIm][Dca]	349	360	[C$_6$MIm][NTf$_2$]	358	358
[C$_3$MIm][NO$_3$]	383	377	[C$_8$MIm][BF$_4$]	371	358
[C$_3$MIm][OAc]	426	416	[C$_8$MIm][Dca]	334	353
[C$_3$MIm][PF$_6$]	352	353	[C$_8$MIm][NTf$_2$]	365	367
[C$_3$MIm][SCN]	369	371	[C$_8$MIm][PF$_6$]	355	348

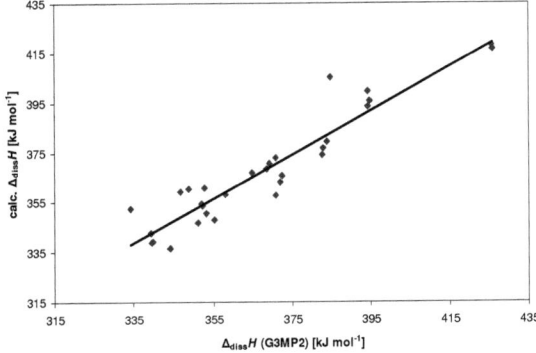

Figure 68: Calculated dissociation enthalpies (according to G3MP2 and eqn. (44), respectively) in kJ mol⁻¹.

4.6.6. Conclusion

For the first time, the Born-Fajans-Haber cycle of phase transitions of ionic liquids was closed on basis of the available experimental and calculated data. We found new approximations for the standard vaporization, solvation, dissociation, and lattice enthalpies, extending traditional VBT-type equations with measures that can be calculated in silico without the need for any experimental input. In the case of lattice enthalpy, the intrinsic error of our new formula is significantly below the one for classical VBT and is of about the same size as the combined experimental and calculative error. The same formula holds true for the dissociation enthalpy with an error comparable to the G3MP2 one, but discarding of the high computational cost associated with the method itself and the tedious optimization of weakly coordinated ion pairs.

For the vaporization enthalpy, a simple formula was found which, similar to already known ones, includes a surface part. A connection of the solvation enthalpy to the HOMO-LUMO gap was also made; however, it requires the calculation of the ion pair and holds only true for a narrow class of compounds,

To close the BFH circle with energies (ΔU), Gibbs energies (ΔG) or entropies (ΔS) instead of enthalpies, there's too little experimental calibration data available at present. If the exact temperature-dependent entropies were known, one could use this to calculate the melting point, as indicated in our earlier study.[87] On the other hand, the Gibbs enthalpy of fusion of [C_6MIm][NTf$_2$], for example, has a slope of about -77 J mol^{-1} K^{-1} in the range of liquid-crystal coexistence,[178] implying that an error of only 1 kJ mol^{-1} in its calculation would introduce an error in the melting point of about 13 °C. This implies that the errors in the derived formulas for phase transition enthalpies are probably too high for a reliable prediction of the melting point of a wide range of ILs.

However, another possibility opens up in order to obtain phase transition temperatures. Without knowing the absolute values of a bulk enthalpy, nor its change with temperature, the ratio of enthalpy and entropy at the equilibrium point could yield a valuable tool to predict this point for new substances, which will be treated in the following chapter.

4.7. The Melting Point[234]

Out of the many relevant physical IL properties, the probably most important to know in advance is the melting point (T_{fus}). Until now, the most successful way for melting point predictions of ionic liquids is the Quantitative Structure-Property Relationship (QSPR) method, especially in newer approaches combined with neural networks.[94,95,235,236,237,238,239,97,93,99] From large training sets with accurate experimental data, physical properties of related compounds may be predicted. Often, the CODESSA code is used.[240] The correlation between experimental and predicted melting points is quite good: for the best model with the best data set (substituted pyridinium bromides), the standard deviation is 7.63 °C.[241] The narrow restrictions in the choice of compounds are its main drawback.

In recent years, simulations with molecular dynamics (MD) have evolved to study the behavior of ILs. The quality of these simulations strongly depends on the employed force fields. Several groups have tuned them specifically for ILs, while others have modified previously existing ones.[242,243,244,245,246,247,248,249,250,251,252] For example, Alavi and Thompson have used MD simulations to predict the melting temperature of [C$_2$MIm][PF$_6$].[98] This study used constant pressure and performed simulations at various temperatures, using the X-ray crystal structure of [C$_2$MIm][PF$_6$] as a starting point for all calculations.[253] At temperatures above T_{fus}, the simulation should refine from the ordered (solid) starting point to a more disordered (liquid) state. On melting, there is a discontinuous change in the density (which decreases) and intermolecular energy (which increases); the onset of these changes is taken to be T_{fus}. In order to avoid problems associated with "superheating", which leads to predicted melting points that are too high, the lattice-defect (void) induced melting method was used.[254,255,256] This involves introducing voids into the starting structures in order to provide an interface for the heterogeneous nucleation of melting, as a result reducing the free-energy barrier for this process. In the end, the demanding simulation indicated a melting point which is about 43 °C too high. Maginn used a similar model for the two polymorphs of [C$_4$MIm]Cl and obtained a T_{fus} which was between 20 and 55 °C too high.[257] The drawbacks of all MD simulations are the high computational demands and the need for a reasonable starting geometry, often the crystal structure.

In our group, we approximated the melting point on a thermodynamic basis using a simple Born-Fajans-Haber (BFH) cycle which assesses the Gibbs enthalpy of fusion ($\Delta_{fus}G$) as the sum of Gibbs lattice and solvation enthalpies, which could be calculated from a standard VBT model and a COSMO calculation using the experimental dielectric constant.[87] We recently found that this methodology is difficult to use for purely theoretical melting point predictions, since for an exact estimation of $\Delta_{fus}G$, both its intercept and slope have to be known with great accuracy. For example, in [C$_6$MIm][NTf$_2$], this enthalpy has a slope of about −77 J mol^{-1} K^{-1} in the range of solid-liquid coexistence,[178] though, implying that an error of only 1 kJ mol^{-1} in its calculation would introduce an error in the melting point of 13 °C. This means that the error intrinsic to the computational methods already makes the reliable determination of the melting point difficult.

In their seminal studies, Yalkowsky et. al pointed out another simple thermodynamic approach to predict the melting point of a compound (in his case exclusively neutral organic molecules), starting from the Gibbs-Helmholtz equation:[215,216,258]

(45) $\Delta G = \Delta H - T\Delta S$

At the melting point ($T = T_{fus}$), solid and liquid phase are in chemical equilibrium ($\Delta G = 0$); therefore, we can state:

(46) $T_{fus} = \dfrac{\Delta_{fus}H}{\Delta_{fus}S}$

Especially gratifying is the fact that, in contrast to our BFH cycle approach,[87] the absolute values of $\Delta_{fus}H$ and $\Delta_{fus}S$ have not to be known, only their ratio. This suggests simpler pathways for prediction. Yalkowsky et al. found universal ways to estimate melting enthalpies and entropies of neutral organic molecules. According to Yalkowsky, the melting entropy can be estimated from the following, simplified equation:

(47) $\Delta_{fus}S = a \cdot \log \sigma + b \cdot \tau + c$

σ is a symmetry number which bears information about the number of equal possibilities the molecule (excluding hydrogen atoms) can be rotationally oriented in the lattice. In this view only rotational axis C_n or S_n but no mirror planes are counted; usually σ represents the index n of the highest C_n or S_n axis. The second parameter τ describes the conformational freedom of the molecule, deduced from the number of its torsion angles. The predicted and measured values agree within the given level of experimental uncertainty of 12.5 J mol^{-1} K^{-1}.[258] The coefficients a, b and c vary slightly in their publications, depending on the data used.

To calculate the melting enthalpy ($\Delta_{fus}H$), Yalkowsky et al. developed a group contribution method, in its latest iteration using 200 groups and 10 proximity factors.[215] For a collection of more than 2000 organic, non-ionic compounds with experimental melting points in the range between 85.5 and 710.5 K, an average absolute error of 30.1 K was observed. Some issues remain with the details of this procedure:

- Errors both in $\Delta_{fus}H$ and $\Delta_{fus}S$ are multiplied when forming the fraction, which is inherent to this kind of approach.
- The assignment of 200 at times quite complex group factors is non-trivial, hard to automatize and very tedious if done by hand.
- Also, as can be seen in Figure 6 in ref. [215], some classes of compounds may form sub-trends, suggesting that the application of a reasonable partition of the data set would minimize the error.

4.7.1. Background

Ionic liquids (ILs) may be considered as 1:1 mixtures of hybrid organic-inorganic molecular ions. Conformation and orientation of any given ion will depend on the counterion, both in the isotropic liquid and the ordered solid phase. In contrast to neutral molecules, the lattice formation of charged ions is governed by strong Coulomb interactions and only in part by the interactions that rule the lattice formation of neutral molecules (dispersion, H-bonds, π-stacking). Also, the liquid phase of ILs contains large Coulomb contributions. So, *a priori* it was not clear to us, if it would be possible to transfer this approach for melting point predictions to ionic liquids.

The group contribution method used by Yalkowsky et al. to assign enthalpies of fusion is not easy to adapt for ionic liquids. Moreover it is difficult to implement in an automated protocol. Therefore, we investigated an independent way to determine this property. In general, for an estimation of the enthalpies of fusion, a detailed knowledge of the solid as well as the liquid state is needed. While an isotropic liquid is randomly disordered and can be subjected to statistical analysis (e.g. with COSMO-RS[15]) without the need for any measurement, the crystal structure and therefore the interaction of complex molecules in a three dimensional lattice is *a priori* unknown and the explicit solid state structure prediction (including crystal system and space group) is computationally exceedingly expensive.[259] However, in the lattice, molecules and molecular ions have:
- a preferred conformation,
- a preferred orientation towards each other and, thus, short and long range order,
- and – depending on pressure and temperature – a variation of both conformation and orientation, possibly leading to the formation of polymorphs.

Without a – prohibitively expensive – complete lattice simulation, the part which the solid plays in melting is largely unknown. As a simplified approach, we assumed that $\Delta_{fus}H$ is a function of volume-based thermodynamics[129] – which is the closest to a statistical examination of the lattice one could get as of now – and of directed interaction enthalpies which are calculated by COSMO-RS in the liquid state only. If successful, this would allow overcoming the tedious and restricted group contribution method for assessing $\Delta_{fus}H$. To eliminate additional factors of uncertainty and also due to a lack of experimental data, however, we did not attempt to determine the independent values of the melting enthalpies and entropies, but instead investigated the entire expression with an ordinary least-squares fit:

(48) $\Delta_{fus}H - T_{fus} \cdot \Delta_{fus}S = 0$

Rather than trying to establish a new method to treat the entropy of fusion, we oriented ourselves towards the published approach.[215] However, since fraction (46) is reducible, we arbitrarily set coefficient c from eqn. (47) to 1. To transfer this approach from neutral molecules to 1:1 salts, we defined σ to be the geometric mean of the individual symmetry factors of cation (σ^+) and anion (σ^-), i.e.

(49) $\sigma = \sqrt{\sigma^+ \sigma^-}$

Because of their S_4 symmetry, σ^+ is 4 for the symmetrical ammonium cations; for all other cations, $\sigma^+ = 1$. For the S_4-symmetric aluminate and boranate anions, $\sigma^- = 4$. For $[NTf_2]^-$ and related symmetric anions, a form with C_2 symmetry exists; therefore, $\sigma^- = 2$. The $[CTf_3]^-$ anion is C_3-symmetric and thus $\sigma^- = 3$. For the remainder of the anions used in this study, $\sigma^- = 1$.

The factor τ describes the torsional freedom of a molecule or molecular ion. To transfer this quantity from neutral molecules to ionic liquids it is taken as the sum of cationic and anionic torsion angles (τ^+ and τ^-) as defined by:

(50) $\tau = \tau^- + \tau^+$

and

(51) $\tau^\pm = n(SP3) + 0.5\,(n(SP2) + n(rings)) - 1$

In eqn. (51), all atoms in a ring are counted as one entity and the number of rings is denoted by n(rings); for all remaining atoms, n(SP3) is the number of sp³-hybridized atoms, n(SP2) the number of sp²-hybridized atoms. Highly symmetrical end groups like methyl or tert.butyl are not included in the count. A -C(CF$_3$)$_2$(CH$_3$) group, as in the ['HT']$^-$ anion, does not count as rotationally symmetrical and thus increases the τ count by 1. If either τ^+ or τ^- would assume negative values, they are set to zero. For example, the $[C_4MIm]^+$ cation consists of one ring and a chain of three CH$_2$ groups; the terminal CH$_3$ groups are not counted (cf. Figure 69). Therefore, its τ^+ value is 3+(0+1)/2-1 = 2.5.

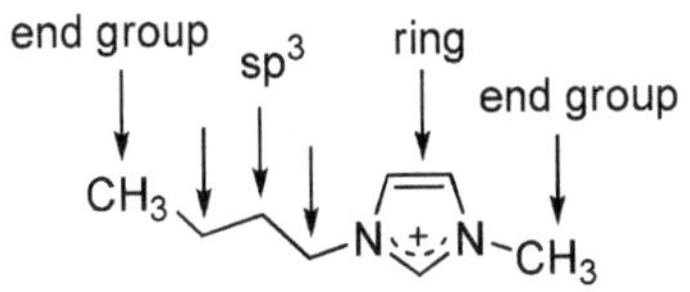

Figure 69: The [C₄MIm]⁺ cation.

The [NTf$_2$]⁻ anion consists of a sp³-hybridized nitrogen atom, two SO$_2$ groups (which don't bear any conformational freedom) and two symmetrical terminal CF$_3$ groups; its τ⁻ value is therefore 1+0+0-1 = 0. This picture, however, is very simplified: due steric hindrance, for example, not all torsional angles do have the same energy profile. Therefore, they are not equally important for the gain of entropy in the melting process. Also, other tautomeric structures may exist, restricting free rotation around an atom (cf. Figure 70), suggesting a more sophisticated approach might be necessary. However, a treatment with statistical thermodynamics would need to consider the entire conformer space for each compound, which is clearly beyond the scope of this study.

Another source of error might be crystal-crystal or mesophasic transitions at lower temperatures which may influence the melting entropy.[260] To minimize this sort of error, we restrained ourselves to ILs with chain lengths < 10 carbon atoms for which the formation of mesophases is much less likely.

Figure 70 also shows that the assigned sp³-hybridization of the N-atom is at least ambiguous if the lower tautomeric form is considered. For the sake of simplicity, we only used the upper tautomeric form (which likely bears most weight, c.f. the N-bound neutral H-NTf$_2$ acid).

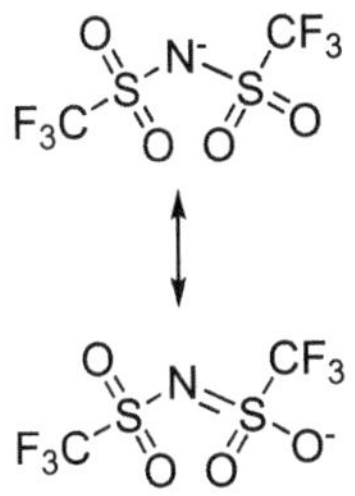

Figure 70: Two of the many tautomeric structures of the [NTf$_2$]⁻ anion.

A list of τ- and σ-values for all ions in this study, along with their volumes, is given in Table 64.

Table 64: Symmetry numbers (σ), torsional freedom numbers (τ), and ionic volumes (V_{ion}, given in nm³) for all treated ions.

ion	σ±	τ±	V_{ion}±	ion	σ±	τ±	V_{ion}±	ion	σ±	τ±	V_{ion}±
[‚BHF']⁻	4	7	0.559	[C₃MPip]⁺	1	1.5	0.209	[N₁,₁,₃,₄]⁺	1	4	0.227
[‚HF']⁻	4	7	0.581	[C₄C₁MIm]⁺	1	2.5	0.218	[N₁,₂,₂,₄]⁺	1	4	0.238
[‚HT']⁻	4	7	0.667	[C₄MIm]⁺	1	2.5	0.197	[N₁,₂,i₃,i₃]⁺	1	2	0.231
[‚PF']⁻	4	3	0.747	[C₄MMorph]⁺	1	2.5	0.222	[N₂,₂,₂,₂]⁺	4	3	0.202
[NTf₂]⁻	2	0	0.233	[C₄MPyr]⁺	1	2.5	0.212	[N₂,₂,₂,₅]⁺	1	6	0.268
[NPf₂]⁻	2	2	0.312	[C₄PIm]⁺	1	4.5	0.256	[N₂,₂,₂,₆]⁺	1	7	0.288
[NNf₂]⁻	2	6	0.463	[C₄Py]⁺	1	2.5	0.191	[N₂,₂,₂,₈]⁺	1	8	0.341
[CTf₃]⁻	3	0	0.335	[C₅MIm]⁺	1	3.5	0.220	[N₂,₂,i₃,i₃]⁺	2	3	0.252
[NFs₂]⁻	2	0	0.148	[C₆C₁MIm]⁺	1	4.5	0.264	[N₃,₃,₃,₃]⁺	4	7	0.310
[NTfaTf]⁻	1	0.5	0.206	[C₆MIm]⁺	1	4.5	0.244	[N₄,₄,₄,₄]⁺	4	11	0.380
[(C₁)₄MIm]⁺	2	0	0.207	[C₁₃MIm]⁺	1	0.5	0.174	[N₄,₄,₄,₆]⁺	1	13	0.437
[(C₂)₂EIm]⁺	1	2.5	0.218	[N₁,₁,₁,₁]⁺	4	0	0.113	[N₅,₅,₅,₅]⁺	4	15	0.486
[AllylMIm]⁺	1	1	0.167	[N₁,₁,₁,₂]⁺	1	0	0.151	[N₆,₆,₆,₆]⁺	4	19	0.576
[C₁MIm]⁺	2	0	0.126	[N₁,₁,₁,₃]⁺	1	1	0.159	[N₇,₇,₇,₇]⁺	4	23	0.666
[C₂C₁MIm]⁺	1	0.5	0.173	[N₁,₁,₁,₄]⁺	1	2	0.183	[N₈,₈,₈,₈]⁺	4	27	0.730
[C₂MIm]⁺	1	0.5	0.156	[N₁,₁,₁,i₃]⁺	1	0	0.172				
[C₃C₁MIm]⁺	1	1.5	0.197	[N₁,₁,₂,₃]⁺	1	2	0.180				

4.7.2. Results

Care has to be taken when establishing computational prediction models, as the parametrization is dependent on the quality of the calibration data. However, melting points of organic salts, as given in literature, may differ noticeably. High impurity content, e.g. organic solvents or moisture, halides, free bases and acids, may dramatically change the melting point by several ten degrees, or even suppress the formation of a lattice, leading to a glassy phase instead.[261,262,263,264] For best results, only melting points of the most pure compounds should be included in the training sets; however, this constraint could in turn limit the validity range of the predictions. Thus, the right balance between size and quality of the training sets is important.

The first set consists of a series of 24 organic salts with aluminate [Al(OC(R)(CF₃)₂)₄]⁻ (R = H (['HF']⁻), CH₃ (['HT']⁻), CF₃ (['PF']⁻))[19,148,19,265,53] and borate [B(OCH(CF₃)₂)₄]⁻ (['BHF']⁻)[266] counterions. Since the isolated compounds were always handled in an inert atmosphere (< 1 ppm H_2O and O_2) and in solvents with < 5 ppm moisture in our group, we are confident that their measured melting points are reliable. Moreover, many of the melting point measurements were reproduced from several independently prepared batches. The aluminates are used since 1999 in our group and are model weakly coordinating anions with little directed interactions in the solid state. Therefore, mainly the variation in the cation is expected to influence the melting process. It should be noted, though, that a highly symmetrical end group like perfluoro-tert.butyl does not add to the conformational freedom and τ^- is only 3 (instead of 7 for the other aluminates/borates). This reduced entropy in ['PF']$^-$ ILs reflected in the low τ^- value is the main reason for their much higher melting points (i.e. the lower gain in entropy upon melting).

In a 0th order approach to obtain the melting points, we attempted to model the part of the enthalpy of fusion completely based on the principles of volume-based thermodynamics.[126,127,128,129] In nucleation and micellization, the molecular volume is a measure for building up order in three spatial dimensions (i.e. Coulomb energy, covalent contributions, dispersion) which is destroyed or weakened when the higher-ordered phase undergoes the transition to the isotropic one.[267,268,269] Thus, we set the cubed molecular radius as the only ordering factor for the melting enthalpy, i.e.

$$(52) \quad T_{fus} / K = \frac{c \cdot r_m^3}{a \cdot \ln \sigma + b \cdot \tau + 1}$$

No y-intercept is needed in the enumerator. With this simple approach, we already obtained an rmse of the calculated melting points of 25.9 °C, where the experimental melting points span a range of 320 °C. Thus, it seems that the molecular volume is the most important thermodynamic ordering principle in phase transition enthalpies, at least for weakly coordinated compounds.[53] Coefficients are found in Table 65, results in Table 66.

In the next stage, we compiled a series of melting points of quaternary alkyl-substituted ammonium ILs – as they are deemed to be model weakly coordinating cations – with the [NTf$_2$]$^-$ anion. In detail, we used [N$_{1,1,1,1}$]$^+$, [N$_{1,1,1,3}$]$^+$, [N$_{1,1,1,4}$]$^+$, [N$_{1,1,2,3}$]$^+$, [N$_{1,1,3,4}$]$^+$, [N$_{1,2,2,4}$]$^+$, [N$_{1,2,i3,i3}$]$^+$, [N$_{2,2,2,2}$]$^+$, [N$_{2,2,2,5}$]$^+$, [N$_{2,2,2,6}$]$^+$, [N$_{2,2,2,8}$]$^+$, [N$_{2,2,i3,i3}$]$^+$, [N$_{3,3,3,3}$]$^+$, [N$_{4,4,4,4}$]$^+$, [N$_{4,4,4,6}$]$^+$, [N$_{5,5,5,5}$]$^+$, [N$_{6,6,6,6}$]$^+$, [N$_{7,7,7,7}$]$^+$, and [N$_{8,8,8,8}$]$^+$; these compounds will be denoted as "set 2". Indeed, we found the same formula to hold true, but with different coefficients (see Table 65). The correlation is slightly worse than for set 1, but still satisfactory (rmse = 27.2 °C, range = 158 °C). Probably, this is due to the different conformations the [NTf$_2$]$^-$ anion can assume and the resulting different coordination and packing in the crystal.[270,271] No special outliers were detected. After we included all known related but less symmetric [NTfaTf]$^-$ compounds (see Figure 71) with cations [N$_{1,1,1,1}$]$^+$, [N$_{1,1,1,2}$]$^+$, [N$_{1,1,1,3}$]$^+$, [N$_{1,1,1,i3}$]$^+$, and [N$_{2,2,2,2}$]$^+$, together with set

2 forming set 2', little changes in the coefficients were obtained; also, the error remained almost constant (see Table 66).

Figure 71, left side: The [NR$^f_2$]⁻ family of anions. R^f = F/ CF$_3$/ C$_2$F$_5$/ C$_4$F$_9$ for [NFs$_2$]⁻, [NTf$_2$]⁻, [NPf$_2$]⁻, and [NNf$_2$]⁻, respectively. Center and right side: The [CTf$_3$]⁻ and [NTfaTf]⁻ anions.

Based upon this, we then applied the same correlation to monocyclic imidazolium compounds with perfluorinated anions related to [NTf$_2$]⁻ (denoted as "set 3"). A larger error was expected, as the anions are varied and the cations differ chemically more than e.g. the ammonium ones. Usually, the hydrogen atom bound to the C2 carbon atom is the most acidic and therefore, the preferred site for anion coordination. However, if it is blocked by an alkyl group, another coordination site has to be found, significantly changing the crystal structure. We investigated all ILs for which a melting point could be gathered from literature, i.e. [Cat][NTf$_2$], where Cat = C$_1$MIm, (C$_1$)$_4$MIm, C$_2$MIm, C$_2$C$_1$MIm, (C$_2$)$_2$EIm, C$_{i3}$MIm, C$_3$C$_1$MIm, C$_4$MIm, C$_5$MIm, C$_6$MIm, C$_6$C$_1$MIm; [C$_2$MIm][A] (where A = NFs$_2$, NPf$_2$, NNf$_2$, NTfaTf, and CTf$_3$), [C$_1$MIm][NNf$_2$], [C$_2$C$_1$MIm][NPf$_2$], and [C$_4$PIm][NNf$_2$] (see Figure 71). We found, rather surprisingly, an error comparable to sets 2 and 2' (rmse = 27.4 °C, range = 135 °C).

Table 65: Coefficients for eqn. (52) used to calculate the melting point. Also given: the range of experimental melting points, the rmse and the average (and maximum) error in °C, abbreviated as err; # = number of compounds in the respective set.

set	1	2	2'	3	2'∪3	1∪2'∪3
a	-0.3952	-0.4043	-0.3653	-0.3585	-0.3889	-0.2978
b [10⁻³]	57.66	61.22	65.69	52.99	57.88	60.58
c [nm⁻³]	559.8	739.8	786.3	682.5	717.8	681.6
#	24	24	19	19	43	67
range	320	158	158	135	165	337
r²	0.9103	0.7346	0.6758	0.5904	0.5118	0.6746
rmse	25.9	27.2	27.4	24.4	32.1	44.9
err	21.2 (66.5)	19.6 (79.0)	23.0 (66.0)	18.6 (53.9)	25.8 (88.5)	36.4 (116)

Table 65 shows that the coefficients for all sets are of comparable size. Therefore, we suspected that a universal formula for all kinds of organic salts might be possible. To investigate this, we combined sets 2' and 3 as well as sets 1, 2', and 3, and included the results in Table 65. The error for each combination increases as more and more different compounds are included. Thus, such a universal formula would only be useful if a huge range of melting points was considered – diminishing the error in comparison – or if more descriptors were added.

We continued to include more directed interactions to the enthalpy of melting that were hitherto neglected in our simple volume-based model by using the COSMO-RS model. The – according to our investigations – best combination for the combined sets 1…3 is given in eqn. (53):

$$(53)\ T_{fus}/K = \frac{c \cdot r_m^3 + d \cdot H_{vdW}^0 + e \cdot H_{ring}}{a \cdot \ln \sigma + b \cdot \tau + 1}$$

Again, no y-intercept is needed in the enumerator. Here, a = -0.3359, b = 0.06867, c = 339.5 nm^{-3}, d = -2.608, e = 7.599 (d, e given in mol kJ^{-1}). We obtained an rmse of 30.5 °C, r^2 = 0.7987 (meaning that the correlation consists of more than 89% linear parts); the average (maximum) error amounts to 24.5 (89.1) °C. The interaction enthalpies (H_{vdW}^0 and H_{ring}) were calculated with COSMO-RS as the sum of the single-ion enthalpies in a 1:1 mixture of cation and anion at 25 °C. Results are found in Table 66 and Figure 72.

Table 66: Experimental and calculated melting points for equations (52) and (53) and the different sets of ILs.

IL	ref.	exp.	eqn. (52), individual sets 1, 2' (2), and 3	eqn. (52), set 1∪2'∪3 (2'∪3)	eqn. (53), set 1∪2'∪3
[AllylMIm]['BHF']	266	43	33.9	73.7	50.1
[AllylMIm]['HF']	266	12	25.9	65.1	44.0
[C$_2$C$_1$MIm]['HF']	266	39.2	37.8	78.3	51.9
[C$_2$MIm]['HF']	266	31.4	20.2	58.4	37.9
[C$_3$MPip]['BHF']	266	106	39.5	80.6	33.5
[C$_3$MPip]['HF']	266	69.4	46.7	88.8	44.3
[C$_4$C$_1$MIm]['HF']	266	0.2	37.9	79.3	48.0
[C$_4$MIm]['HF']	272	39.9	24.9	64.5	37.0
[C$_4$MIm]['PF']	148	149	149.8	203.6	160.6
[C$_4$MMorph]['HF']	266	31.4	40.1	81.8	38.1
[C$_4$MPyr]['BHF']	266	55	94.1	140.8	98.9
[C$_4$MPyr]['HF']	272	50	34.5	75.4	37.2
[C$_4$MPyr]['PF']	148	202	162.7	218.2	161.4
[C$_4$Py]['HF']	266	36.1	21.6	60.7	30.7
[C$_6$MIm]['HF']	266	4.9	25.8	66.1	35.5
[N$_{1,1,1,1}$]['HT']	265	96	117.5	129.4	139.9
[N$_{1,1,1,1}$]['PF']	265	320	302.4	296.7	320.3
[N$_{1,1,1,1}$]['HF']	265	61	84.6	95.5	108.0
[N$_{2,2,2,2}$]['HF']	265	56	97.8	116.4	102.5
[N$_{2,2,2,2}$]['HT']	265	111	129.4	149.5	132.7
[N$_{2,2,2,2}$]['PF']	265	308	282.5	294.9	271.5
[N$_{4,4,4,4}$]['HF']	265	42	71.1	99.2	74.5

Compound	Ref				
[N$_{4,4,4,4}$]['HT']	265	108	97.6	127.8	99.1
[N$_{4,4,4,4}$]['PF']	265	199	193.6	225.5	179.9
[N$_{1,1,1,1}$][NTf$_2$]	273	133	129.0 (131.7)	40.1 (109.1)	118.0
[N$_{1,1,1,i3}$][NTfaTf]	276	64	25.2	-28.4 (6.5)	24.3
[N$_{1,1,1,1}$][NTfaTf]	276	15	-15.6	-49.3 (-37.1)	-21.3
[N$_{1,1,1,2}$][NTfaTf]	276	19	36.4 (23.8)	-9.9 (14.3)	30.9
[N$_{2,2,2,2}$][NTf$_2$]	273,274,275	10	-25.5	-57.0 (-44.7)	-26.0
[N$_{2,2,2,2}$][NTfaTf]	276	17.5	35.8 (24.2)	-8.8 (15.7)	33.4
[N$_{1,1,1,3}$][NTf$_2$]	276	21	0.8	-35.1 (-22.1)	-7.8
[N$_{1,1,1,3}$][NTfaTf]	276	-10	34.3 (22.7)	-10.1 (14.3)	25.6
[N$_{1,1,2,3}$][NTf$_2$]	278	20	31.0 (21.2)	-10.2 (14.5)	20.7
[N$_{3,3,3,3}$][NTf$_2$]	273	9	38.1 (28.1)	-4.0 (21.3)	20.7
[N$_{1,1,1,4}$][NTf$_2$]	229,278	140	74.0 (61.0)	23.9 (51.5)	50.9
[N$_{1,2,2,4}$][NTf$_2$]	277	106	123.6 (126.5)	49.1 (111.6)	95.5
[N$_{1,1,3,4}$][NTf$_2$]	278	20	38.2	-11.0 (24.4)	23.0
[N$_{4,4,4,4}$][NTf$_2$]	273,274,275	0	24.9 (16.5)	-13.3 (11.3)	17.0
[N$_{2,2,2,5}$][NTf$_2$]	87	20	21.5 (13.7)	-15.3 (9.2)	10.1
[N$_{5,5,5,5}$][NTf$_2$]	274	14	17.6 (10.8)	-17.2 (7.3)	8.4
[N$_{2,2,2,6}$][NTf$_2$]	273	148	113.0 (106.4)	50.8 (94.9)	80.9
[N$_{4,4,4,6}$][NTf$_2$]	273	105	101.9 (104.8)	41.9 (96.3)	72.2
[N$_{6,6,6,6}$][NTf$_2$]	274	92.3	68.3 (71.1)	19.8 (66.5)	53.0
[N$_{7,7,7,7}$][NTf$_2$]	274	26	9.0 (3.8)	-22.3 (1.9)	3.8
[N$_{2,2,2,8}$][NTf$_2$]	278	25.2	48.9 (51.6)	7.2 (49.3)	41.1
[N$_{8,8,8,8}$][NTf$_2$]	274	-6.8	31.7 (34.3)	-5.0 (33.6)	31.4
[N$_{1,2,i3,i3}$][NTf$_2$]	273	11.2	17.6 (20.2)	-15.3 (20.5)	23.7
[N$_{2,2,i3,i3}$][NTf$_2$]	273	31.2	8.8 (11.3)	-21.6 (12.4)	18.7
[(C$_1$)$_4$MIm][NTf$_2$]	275	118	108.6	87.9 (140.0)	110.0
[(C$_2$)$_2$EIm][NTf$_2$]	271	57	18.9	7.4 (33.5)	18.3
[C$_1$MIm][NNf$_2$]	279	69	42.2	18.0 (56.0)	26.7
[C$_1$MIm][NTf$_2$]	279	26	28.6	12.2 (53.4)	47.3
[C$_2$C$_1$MIm][NPf$_2$]	275	25	31.4	19.3 (46.6)	24.0
[C$_2$C$_1$MIm][NTf$_2$]	176,275	26	18.1	9.9 (36.0)	26.0
[C$_2$MIm][CTf$_3$]	275	39	87.9	72.0 (113.2)	92.4
[C$_2$MIm][NFs$_2$]	280	-13	-58.8	-64.8 (-45.7)	-37.0
[C$_2$MIm][NNf$_2$]	279	31	22.9	6.4 (32.9)	7.5
[C$_2$MIm][NPf$_2$]	275	-1	12.0	0.7 (26.2)	9.0
[C$_2$MIm][NTf$_2$]	87,275	-17	-1.4	-9.0 (15.3)	14.3
[C$_2$MIm][NTfaTf]	276	-1.5	-55.4	-57.2 (-45.2)	-45.6
[C$_3$C$_1$MIm][NTf$_2$]	218,275	13.2	19.7	9.7 (35.9)	23.7
[C$_4$MIm][NTf$_2$]	87	-5	4.3	-6.6 (18.2)	12.3
[C$_4$PIm][NNf$_2$]	281	36.5	44.9	23.6 (52.1)	10.9
[C$_5$MIm][NTf$_2$]	87	-10	5.3	-7.0 (17.9)	11.3
[C$_6$C$_1$MIm][NTf$_2$]	177	-5.1	17.8	3.7 (29.7)	16.4
[C$_6$MIm][NTf$_2$]	229	-6	5.7	-7.8 (17.2)	9.9
[^{i}C$_3$MIm][NTf$_2$]	275	16	18.7	10.5 (36.6)	28.1

The local order of a liquid permits an optimized fit of the ions compared to the lattice.[282] Rings tend to pack better and thus will give more close contacts than non-ring containing molecules; therefore, e·H_{ring} is < 0 and d·H_{vdW}^0 is > 0. The ring enthalpy is only dependent on the size of the ring, not on its composition (and it's also independent from temperature). All quantities necessary to calculate the melting point are given in Table 67.

Table 67: Miscellaneous quantities necessary for the prediction of the melting point of the investigated ILs. All energies are given in kJ mol^{-1}, other units are given in square brackets.

IL	H_{vdW}^0	H_{ring}	IL	H_{vdW}^0	H_{ring}
[AllylMIm]['BHF']	-85.9	-4.09	$[N_{1,1,1,1}][NTf_2]$	-56.3	0.00
[AllylMIm]['HF']	-89.4	-4.09	$[N_{1,1,1,1}][NTfaTf]$	-52.9	0.00
$[C_2C_1MIm]['HF']$	-88.0	-4.09	$[N_{1,1,1,2}][NTfaTf]$	-55.8	0.00
$[C_2MIm]['HF']$	-85.8	-4.09	$[N_{1,1,1,3}][NTf_2]$	-62.9	0.00
$[C_3MPip]['BHF']$	-84.7	-4.90	$[N_{1,1,1,3}][NTfaTf]$	-59.5	0.00
$[C_3MPip]['HF']$	-88.3	-4.90	$[N_{1,1,1,4}][NTf_2]$	-68.6	0.00
$[C_4C_1MIm]['HF']$	-94.6	-4.09	$[N_{1,1,1,3}][NTfaTf]$	-58.4	0.00
$[C_4MIm]['HF']$	-92.5	-4.09	$[N_{1,1,2,3}][NTf_2]$	-65.8	0.00
$[C_4MIm]['PF']$	-99.7	-4.09	$[N_{1,1,3,4}][NTf_2]$	-73.3	0.00
$[C_4MMorph]['HF']$	-90.9	-4.90	$[N_{1,2,2,4}][NTf_2]$	-72.0	0.00
$[C_4MPyr]['BHF']$	-86.2	-4.09	$[N_{1,2,i3,i3}][NTf_2]$	-69.1	0.00
$[C_4MPyr]['HF']$	-89.7	-4.09	$[N_{2,2,2,2}][NTf_2]$	-67.4	0.00
$[C_4MPyr]['PF']$	-96.9	-4.09	$[N_{2,2,2,2}][NTfaTf]$	-64.1	0.00
$[C_4Py]['HF']$	-92.4	-4.90	$[N_{2,2,2,5}][NTf_2]$	-81.6	0.00
$[C_6MIm]['HF']$	-99.6	-4.09	$[N_{2,2,2,6}][NTf_2]$	-83.1	0.00
$[N_{1,1,1,1}]['HF']$	-77.2	0.00	$[N_{2,2,2,8}][NTf_2]$	-91.6	0.00
$[N_{1,1,1,1}]['HT']$	-83.0	0.00	$[N_{2,2,i3,i3}][NTf_2]$	-71.8	0.00
$[N_{1,1,1,1}]['PF']$	-84.8	0.00	$[N_{3,3,3,3}][NTf_2]$	-82.8	0.00
$[N_{2,2,2,2}]['HF']$	-87.1	0.00	$[N_{4,4,4,4}][NTf_2]$	-100	0.00
$[N_{2,2,2,2}]['HT']$	-93.7	0.00	$[N_{4,4,4,6}][NTf_2]$	-108	0.00
$[N_{2,2,2,2}]['PF']$	-94.5	0.00	$[N_{5,5,5,5}][NTf_2]$	-117	0.00
$[N_{4,4,4,4}]['HF']$	-117	0.00	$[N_{6,6,6,6}][NTf_2]$	-134	0.00
$[N_{4,4,4,4}]['HT']$	-124	0.00	$[N_{7,7,7,7}][NTf_2]$	-151	0.00
$[N_{4,4,4,4}]['PF']$	-123	0.00	$[N_{8,8,8,8}][NTf_2]$	-169	0.00
$[(C_1)_4MIm][NTf_2]$	-69.9	-4.09	$[C_2MIm][NTf_2]$	-66.3	-4.09
$[(C_2)_2EIm][NTf_2]$	-73.7	-4.09	$[C_2MIm][NTfaTf]$	-61.4	-4.09
$[C_1MIm][NNf_2]$	-83.2	-4.09	$[C_3C_1MIm][NTf_2]$	-70.9	-4.09
$[C_1MIm][NTf_2]$	-62.9	-4.09	$[C_4MIm][NTf_2]$	-74.1	-4.09
$[C_2C_1MIm][NPf_2]$	-73.6	-4.09	$[C_4PIm][NNf_2]$	-99.8	-4.09
$[C_2C_1MIm][NTf_2]$	-67.1	-4.09	$[C_5MIm][NTf_2]$	-78.1	-4.09
$[C_2MIm][CTf_3]$	-73.9	-4.09	$[C_6C_1MIm][NTf_2]$	-82.5	-4.09
$[C_2MIm][NFs_2]$	-58.2	-4.09	$[C_6MIm][NTf_2]$	-82.1	-4.09
$[C_2MIm][NNf_2]$	-86.1	-4.09	$[^iC_3MIm][NTf_2]$	-67.7	-4.09
$[C_2MIm][NPf_2]$	-71.2	-4.09			

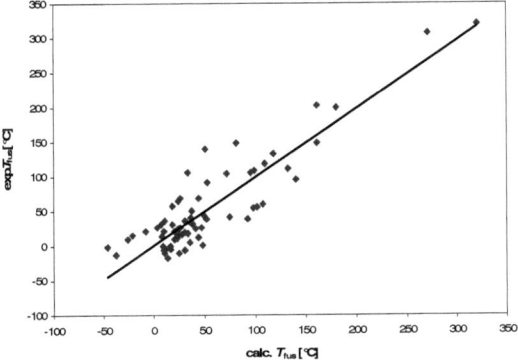

Figure 72 : A plot of experimental vs. calculated (according to eqn. (53)) melting points.

4.7.3. Conclusion

For a multitude of common ionic liquids with diverse ions, our melting point predictions gave remarkable results, which shows that the adaptation and further refinement of the method proposed by Yalkowsky et al. is possible. Its quality is comparable to today's QSPR methods, but has a broader basis in fundamental thermodynamics and needs much less specialized descriptors. The largest difficulty as of now, it seems, is the lack of knowledge of the configuration in the lattice, and thus the absolute magnitude of the directed interactions released upon melting. Unlike molecular substances, organic salts are 1:1 mixtures of anion and cation. The structuring of both their solid and liquid phase is governed not only by dispersive, H-bonding or other directed interactions, but also by Coulomb interaction. Anions and cations may alter their configuration to better fit each other. Also, in the liquid, there will be an ensemble of conformers, whereas we only investigated the most stable one and conjectured that its liquid interaction enthalpies are representative of the difference of interactions in the lattice and in the liquid phase.

Equation (54), which includes all 67 ILs used in this study, gives rather encouraging results: For a 339 °C range of melting points, rmse = 30.5 °C, while using only three descriptors to model the part of the melting enthalpy instead of two hundred as in ref. 215. This suggests that future improvements might expand our model to suit most organic 1:1 salts encountered in real life.

4.8. The Glass Transition Temperature

4.8.1. Background

On melting a crystal, there is a jump in enthalpy, equivalent to a singularity in C_p, which indicates a phase transition of the first order (see Figure 73). For vitrification, which occurs on cooling, however, the picture is not as clear; if the glass transition is a true phase transition of second or even first order is still heavily disputed. In any case, there's a very flat energy hypersurface for the position of the particles in a glass former relative to each other. The voids in the liquid state are, according to MD simulations, Boltzmann-distributed and may have arbitrary sizes,[283] so a certain mobility of the ions may always be expected. While cooling the liquid, the fluidity nears zero, meaning there is no macroscopic motion observable over very long timescales. This is caused by the lack of (thermal) energy needed for wandering from one local minimum to another as well as for surface formation, both of which would be needed for the rearrangement of molecules into a lattice. Thus, below a certain temperature, particles may get trapped at the local minimum they are in at the moment, leading to vitrification. At heating, relaxation of the glass to a regular lattice may occur, which is called "cold crystallization".[284] This is because even little differences in entropy between glass and crystal are multiplied with the temperature, meaning continuously more thermal energy is at disposal which can be used for lattice formation. On further warming, these crystals show a sharp melting point.

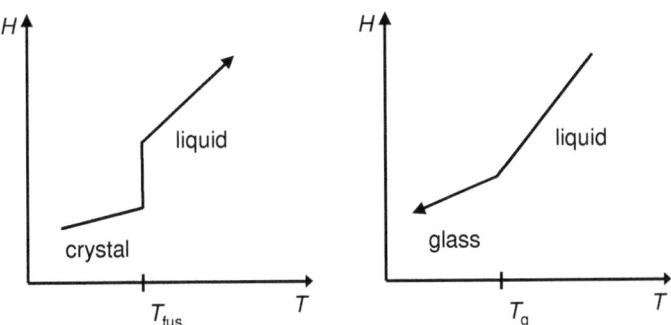

Figure 73: The different phase behavior of a crystal on melting and a vitrificating liquid on freezing.

It was shown that the phase transition behavior of ILs strongly depends on the absence or presence of crystal seeds and their size. In the same substance, few, small seeds are more likely to lead to a glass transition; more and larger seeds induce crystallization.[263] This may be due to surface energy, which helps in the formation of a phase frontier that large crystal seeds may bring; smaller seeds may have too little an influence or even bring more disorder. Other than that, from a purely physical point of view, vitrification is preferred, if

- very different particle sizes are present (which is usually not the case in ionic liquids: the volume ratio of cation and anion rarely exceeds 2);
- particles have little mobility even at elevated temperatures;
- viscosity decreases rapidly with temperature;
- thermal conductivity is high.

The vitrification temperature of ILs, alkali metal salts and polymers was connected in a loose and non-linear fashion to the molar volume;[285] however, especially for certain orthoborate ILs, no such relation exists.[286] In polymers and inorganic glasses, T_g is also linearly connected to the microhardness.[287] The branching of alkyl chains increases T_g in all observed cases,[288] which is the same behavior as for T_{fus}; it can be attributed to a better packing in the solid state.

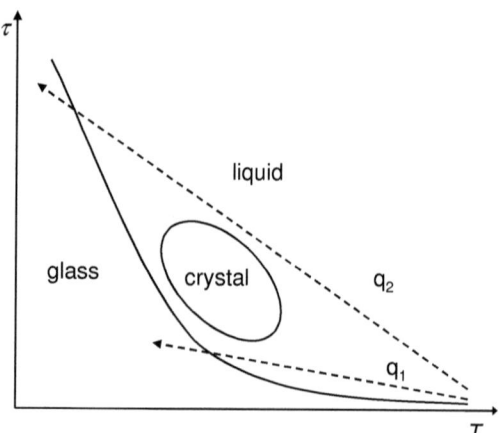

Figure 74: The phase behavior of ionic liquids in dependence on temperature and relaxation time (schematical).

The cooling rate is generally defined as:

(54) $q = -\dfrac{\partial T}{\partial t}$

where T = temperature, t = time. Derived from Figure 2 in ref. [289], Figure 74 is a sketch of the relations an ionic liquid undergoes on cooling. For the cooling rates, $q_1 > q_2$, i.e. at very fast and very slow cooling, a glass will form. The area in which crystallization occurs lies in between; however, depending on the IL in question and on the impurities (crystallization nuclei) present, its position and size can vary considerably. Additionally, it is possible for an IL to show no crystallization at all upon cooling. Obviously, this is a simplified picture, as it is missing entropy as a third dimension. Once a system is "trapped" in the crystalline entropy minimum, it will never return to the liquid or glassy state at cooling. Also, on warming, the diagram will look different as heat ($\Delta Q = T \cdot \Delta S$) is continuously added.

Some ions seem to promote vitrification more than others. For example, dry RTILs containing $[BF_4]^-$, $[PF_6]^-$ or substituted pyridinium cations very often form glasses, i.e. the crystal region in Figure 74 vanishes. Both $[BF_4]^-$ and $[PF_6]^-$ are chaotropes, i.e. they interrupt the structuring of water,[290] or, more generally put, any strongly hydrogen-bound network. However, long alkyl chains shift the behavior of ILs to kosmotropicity, which is contrary to the notion that they encourage vitrification. Probably there's a superimposed effect because of van-der-Waals interactions, which play a negligible role in water but not in ILs.

Usually, T_g will be lower, the lower q; however, we did not find any literature record systematically measuring this connection for ILs.

Table 68: Literature-known vitrification temperatures T_g [°C] of some ILs.

IL	T_g [°C] (ppm water)				
$[C_2MIm][C_2SO_4]$	-86.2 (180)[225]	-78.4 (2000)[291]			
$[C_2MIm][NTf_2]$	-87 (40)[a,229]	-92 (230)[176]			
$[C_4MIm][BF_4]$	-83 (40)[229]	-85 (50)[227]	-87.4 (800)[218]	-85 (1900)[176]	-97 (4530)[292]
$[C_4MIm][NTf_2]$	-87 (40)[229]	-86 (460)[176]	-104 (474)[292]	-102 (3280)[292]	
$[C_4MIm][OTfa]$	-86 (12)[171]	-78 (40)[229]			
$[C_4MIm][PF_6]$	-77 (40)[229]	-80 (590)[292]	-83 (11700)[292]		
$[C_6MIm][NTf_2]$	-88.9 (5)[178]	-84.2 (31)[177]			
$[C_6MIm][PF_6]$	-78 (472)[292]	-75 (8837)[292]			
$[C_8MIm][PF_6]$	-82 (388)[292]	-75 (6666)[292]			

[a] Only observed at rapid cooling.

Table 5 in ref. [176] gives an overview of different glass temperatures from literature. Additionally, as Table 68 shows, the vitrification point of hydrophilic ILs (i.e. with anions $[C_2SO_4]^-$ and $[OTfa]^-$) generally seems to depend more strongly on water content than the hydrophobic ones. The rather large experimental uncertainty in determining T_g and the low range of values (T_g is within 29 °C) need to be taken into account as well. The third and forth values for $[C_4MIm][NTf_2]$ seem to be exceptions since they are much below usually observed vitrification points for ILs; also, they do not fit with the homologue [C#MIm][NTf$_2$] ILs (with # = 2, 6). Based on these observations, we think that for ILs with < 500 ppm water, the change in water content is smaller than the experimental uncertainty. At this point, there is about 1 water molecule in 120 molecules of a typical IL (M = 300 g mol^{-1}). In the following, we will be using the T_g values of the ILs with the lowest water content as the reference values.

The question is whether there are finer trends observable. For polymers it is known that the length of the residues (side groups) can both increase and decrease T_g.[293] A decreased value can be attributed to higher chemical interaction between the side groups, an increased value to higher flexibility of the entire molecule (which is connected to more available free volume). If we look at all known dry (< 500 ppm moisture) ionic liquids with a vitrification temperature T_g < -85 °C, we find them to consist exclusively of imidazolium compounds with rather short alkyl chains ([C#MIm][A]; see Table 69).

Table 69: Anions of imidazolium ILs with T_g < -85 °C (# = number of carbons in the longest alkyl chain).[171,225,227,229,294]

#	[C#MIm]$^+$	[C#EIm]$^+$
2	[BF$_4$]$^-$, [C$_2$SO$_4$]$^-$,	
3	[BF$_4$]$^-$	
4	[NTf$_2$]$^-$, [OTfa]$^-$	[NTf$_2$]$^-$
6		[NTf$_2$]$^-$

A very different picture arises if we look at dry ILs with T_g > -70 °C: they consist mostly of [Cat][NTf$_2$] compounds shown in Table 70 which contain longer alkyl chains and are either functionalized or highly substituted. [C$_2$MIm][OTos] and [C$_2$EIm][NTf$_2$] don't fit well into this scheme at first sight.[225,294] However, the tosylate anion could be considered being functionalized as well, since it may form directed π-π interactions in the liquid. [C$_2$EIm]$^+$ can assume C_2 symmetry. As a high symmetry usually introduces higher melting points (because of a lower entropy of fusion; see section 4.7), it is reasonable to suggest that symmetry also somewhat influences vitrification.

Table 70: [Cat][NTf$_2$] ILs with T_g > -70 °C.[177]

[Cat]$^+$	functionalization	number of substituents
[(C$_4$)$_2$Nic]$^+$	carboxylate	2
[C$_6$DmaPy]$^+$	amino	3
[C$_6$C$_2$(C$_1$)$_2$Py]$^+$	-/-	4
[C$_6$C$_3$(C$_2$)$_2$Py]$^+$	-/-	4

4.8.2. Results

The Kauzmann-Beaman rule gives a relationship between melting and vitrification, i.e. $T_{fus} \approx a \cdot T_g$, where a is usually given as approximately 1.5.[295] For this and further purposes, we built two sets of ILs. The first set contains all dry ILs for which T_{fus} as well as T_g are known, and consists of [Cat][NTf$_2$], with Cat = N$_{1,1,1,4}$, C$_4$MPyr, C$_2$EIm, C$_4$MIm, C$_6$MIm, C$_6$Py, C$_6$(C$_1$)$_2$Py, C$_6$C$_1$DmaPy, (C$_4$)$_2$Nic; [C$_2$MIm][A], with A = BF$_4$, OTos; [C$_4$MIm][A], with A = PF$_6$, OTfa; and [C$_3$MIm][BF$_4$]. For comparison, [C$_2$MIm][NTf$_2$], [C$_4$MIm][OTf] and [C$_4$Py][NTf$_2$] will form set 2; they were described as only showing vitrification at rapid cooling.[229]

To adapt the formula, we calculated a best fit from all ILs of set 1 and obtained:

(55) $T_g = 0.256 \cdot T_{fus} + 124$

However, we obtained an rmse of 8.3 °C for set 1 (see Table 71), which seems too high for the small range of T_g values of 33.9 °C. Additionally, r² is only 0.1905, meaning that important parts are missing in this correlation. For an explanation, looking only at dry imidazolium [NTf$_2$]⁻ ILs with different chain lengths, we can see that the melting points display a higher variation than the glass transition points (Figure 75). This is mirrored, for example, in the [C$_2$MIm]⁺ and [C$_4$MIm]⁺ ILs from Table 71 where different anions impose comparatively large changes in T_{fus}, but not in T_g, meaning that vitrification is governed by different principles than crystallization. In a crystal, the packing depends strongly on hydrogen bonds and chain length,[296,297] as well as on molecular configuration and other factors; in the glassy phase, however, there would be a statistical mixture of all possible orientations of molecules toward each other, somewhat blurring the differences.

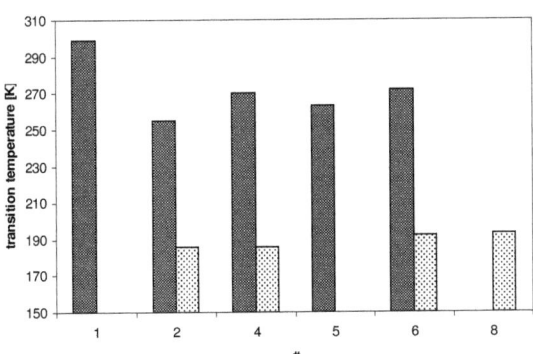

Figure 75: Melting (dark columns) and vitrification (lighter columns) temperatures for [C$_\#$MIm][NTf$_2$] ionic liquids.[87,229,298]

Table 71: Experimental melting and vitrification points and calculated values (according to eqn. (55)) for dry ILs of sets 1.

IL	exp. T_{fus} [°C]	exp. T_g [°C]	calc. T_g [°C]
[C_2MIm][BF_4]	13.0[227]	-92.1[227]	-75.4
[C_3MIm][BF_4]	-17.1[227]	-88.1[227]	-83.1
[C_4MIm][NTf_2]	-3.0[87]	-87.1[229]	-79.5
[C_4MIm][OTfa]	23.3[171]	-86.1[229]	-72.7
[C_4MPyr][NTf_2]	-15.1[87]	-83.1[229]	-82.5
[C_6MIm][NTf_2]	-1.0[298]	-81.1[177]	-79.0
[C_6Py][NTf_2]	-0.1[177]	-77.2[177]	-78.7
[C_4MIm][PF_6]	10.0[87]	-77.1[229]	-76.1
[$C_6(C_1)_2$Py][NTf_2]	9.9[177]	-76.2[177]	-76.2
[$N_{1,1,1,4}$][NTf_2]	19.0[229]	-74.1[229]	-73.8
[C_6C_1DmaPy][NTf_2]	-2.1[227]	-72.2[177]	-79.2
[C_2EIm][NTf_2]	-10.6[294]	-67.3[294]	-81.4
[C_2MIm][OTos]	49.8[222]	-59.2[225]	-65.9
[$(C_4)_2$Nic][NTf_2]	14.9[177]	-58.2[177]	-74.9

Alternatively, a simple, universal connection to the linear expansion coefficient was suggested for all glass-forming liquids in a temperature range of more than 1350 °C:[299]

(56) $\alpha_L = a/T_g + b$

For isotropic solids like liquids or glasses, $\alpha_p \approx 3 \cdot \alpha_L$, so there should also be a connection to the cubic thermal expansion coefficient as calculated in eqn. (11) or (16); reformulated, we get:

(57) $T_g = a/(\alpha_p - b)$

Probably, there's much merit in restricting oneself to a certain class of compounds in order to eliminate additional degrees of chemo-physical freedom. The largest subset of compounds with < 500 ppm moisture for which a vitrification temperature could be gthered from literature is formed by pyridinium ILs with the [NTf_2]$^-$ anion. Thus, we built a new set, called "set 2", consisting of 10 ILs which we used for calibration,[177] i.e. [C_4C_1Py]$^+$, [$(C_4)_2$Nic]$^+$, [C_6Py]$^+$, [C_6C_1Py]$^+$, [C_6DmaPy]$^+$, [$C_6(C_1)_2$Py]$^+$, [C_6C_1DmaPy]$^+$, [$C_6C_2(C_1)_2$Py]$^+$, [$C_6C_3(C_2)_2$Py]$^+$, and [C_8C_1Py]$^+$ (for structures, see Table 2), and obtained coefficients a and b from a linear fit of the calculated α_p and T_g. All results are given in Table 72.

Different from ref. [299], the slope a in eqn. (56) is negative, which implies an unknown physical significance. Also, the smallest calculated α_p (according to eqn. (16)) in set 2 is 0.0000658 K^{-1}, whereas the largest α_p in ref. [299] is less than a third of this value. Thus, it could either be that the relation presented in eqn. (56) changes when moving to larger α_p, or when moving to ionic liquids in general.

Table 72: Coefficients and errors (in °C) in the prediction of T_g as used in eqn. (57), with α_p calculated according to the equations given in the first column.

eqn.	a	b	r^2	rmse
(11)	-0.117	0.00128 K^{-1}	0.5338	4.0
(16)	-0.0537	0.000942 K^{-1}	0.6121	3.6

If we calculate T_g for all given compounds and compare them with their experimental counterparts, we obtain Table 74 and Figure 76 (shown only for the case of eqn. (16)). The range of vitrification points is 26 °C.

The vitrification temperature is generally dependent on the cooling rate, as has been known for a long time.[300,301] The Bartenev equation states that:[300]

(58) $1/T_g = C_1 - C_2 \ln q$

where q is the cooling rate (in K s^{-1}); C_1 and C_2 are substance-specific constants, with $C_2/C_1 \approx 0.03$. For different organic and inorganic glasses, C_1 is roughly between 1 and $3 \cdot 10^{-3}$ K^{-1}, whilst C_2 is between 3 and $10 \cdot 10^{-5}$ K^{-1}.[302,303] From equations (56) and (58), a correlation between cooling rate and expansion coefficient follows:

(59) $\alpha_L = aC_1 \cdot (1 - 0.03 \ln q) + b$

This is a purely kinetic effect and, since both equations are multiplied, expected to be very small, so only very precise measurements could yield an experimental proof.

For an even more accurate calculation of T_g, we investigated a different approach. In the equilibrated state, which takes place with sufficient slow cooling, we can write:

(60) $T_g = \Delta_{vitr}H/\Delta_{vitr}S$

The vitrification point and therefore, according to eqn. (55), the ratio $\Delta_{vitr}H/\Delta_{vitr}S$, is in the same order of magnitude as the melting point. There is no discontinuous change of the liquid enthalpy at the vitrification point; there is, however, still the change of enthalpy with temperature, which is expected to be rather small (for comparison: the melting enthalpy, according to Sidgwick, would only change about 1 kJ mol^{-1} per 18 °C).[217] We can now calculate the ratio in eqn. (60) without necessarily knowing the size of either enumerator or denominator, using the method from Yalkowsky et al.[215,216,258] In this approach, the melting entropy is expressed by a combination of conformational freedom and a symmetry factor.

Neither is expected to take a dramatic turn at the glass transition temperature. Therefore, in a first approximation, it seems reasonable to try to describe $\Delta_{vitr}S$ – as the difference of liquid and glass-state entropy – with volume-based thermodynamics (see eqn. (25) for the general case of an IL at 25 °C). Thus, we get:

(61) $\Delta_{vitr}S = a \cdot r_m^3 + b$

Also, we can describe the change of the liquid enthalpy with temperature as a function of the standard interaction energy (H_l^0) of an ionic liquid at some temperature near the range of vitrification temperatures, assuming its increase is linear in this range:

(62) $\Delta_{vitr}H = \dfrac{\partial H_l}{\partial T} = c \cdot H_l^0 + d$

It has to be noted, though, that the fraction in eqn. (60) is reducible; the real size of the coefficients is therefore impossible to determine.

If we combine equations (60) to (62) and reduce, we get:

(63) $T_g = \dfrac{e \cdot H_l^0 + f}{g \cdot r_m^3 + 1}$

As noted above, the constituents of the enthalpy are system-specific, i.e. best results are achieved if one ion is held constant and the other is varied in a specific range. Thus, we investigated set 2 and found that H_l^0 can be excellently estimated with specific interaction energies – in this case, the hydrogen bonding and misfit enthalpies as given by COSMO-RS[15] for the 1:1 mixture of cation and anion – scaled by the solvent-accessible surface (Ŝ), leading to the following equation:

(64) $T_g = \dfrac{\dfrac{e_1 \cdot H_{HB}^0 + e_2 \cdot H_{MF}^0}{\hat{S}} + f}{g \cdot r_m^3 + 1}$

Here, we get an rmse of 2.2 °C, with $r^2 = 0.9215$. We did not find any other reasonable descriptor to significantly further lower the error. Thus, we stuck with the simpler solution; best-fit analysis to the calibration set yielded the coefficients listed in Table 73.

Table 73: Coefficients for eqn. (64).

coefficient	e_1	e_2	f	g
value	-10.5	40.6	-193	-1.17
unit	nm^2 K mol kJ^{-1}		K	nm^{-3}

Table 74 lists the results. We need to consider that the investigated ionic liquids are chemically quite diverse and often functionalized. The sign of g needs to be reversed for $\Delta_{vitr}S$ to increase with r_m, which is physically reasonable. Then, $\Delta_{vitr}S$ is always < 0, which would be expected when going from the liquid to the glass at lower temperature. This also means that the signs of coefficients e to f need to be reversed as well. Then we see that the enthalpy decrease is governed by $e_1 \cdot H_{HB}^0$ and $e_2 \cdot H_{MF}^0$, which are both < 0. In fact, they do not denominate absolute enthalpies, but, according to eqn. (62), the slope of the enthalpy at cooling, which is negative.

Obviously, for very large systems, the formula cannot hold true, as for r_m increasing to 0.95 nm, T_g would approach infinity. The investigated r_m, though, are all in the range 0.43 to 0.59 nm.

In eqn. (64), $[(C_4)_2Nic]^+$ forms a rather big exception, probably due to the completely missing description of functionalization in the formula. Two ILs not present in the calibration set 2, i.e. with cations $[C_4Py]^+$ and $[C_6MPipPy]^+$, form the largest overall errors. In the first case, it needs to be noted that both eqn. (57) and (64) predict a much lower vitrification point, consistent with the picture that the measured T_g could only be obtained at rapid cooling (see also Figure 74). In the second case, the water content is unknown. Due to its functionalization, the physical behavior of this IL may be quite susceptible to higher moisture content. It is also possible that cations containing more than one ring system have less conformational freedom and achieve better packing, making jamming occur at higher temperatures.

Additionally, it needs to be noted that only the most stable conformer of any ion was investigated with COSMO-RS; if the entire conformer space – present in the liquid and glassy state – would be included, interaction enthalpies would slightly change (cf. section 2.1.4).

From the results collected in Table 74, it can be observed that the majority of ILs with short chains (with about one through four carbon atoms) forms crystals on cooling. Cations with long, straight alkyl chains (with usually at least 10 carbon atoms) aggregate in the lattice, i.e. their structural arrangement is governed by large dispersive van-der-Waals interactions.[304] This type of ILs also strongly tends to form liquid crystals.[159] Thus it seems that on cooling, glass formers are kinetically trapped in a disordered state: they are not Coulomb-ordered enough to form crystals and not vdW-ordered enough to form liquid crystals. So it can be noted that, in some respect, ILs can behave like polymers: They may form glasses even at very slow cooling rates, and they may show unusual phase behavior, especially "cold crystallization".[284] More generally put, it can be stated that ILs have a certain tendency to form "soft matter", which, apart from mesophases (liquid and plastic crystals)[305,306,307,308] also includes micelles, which will be treated in the following section.

Table 74: Overview of experimental and calculated (with eqn. (57) and (64) for (I) and (II), respectively) T_g (given in °C). The first ten entries form the calibration set.

cation	α_p [10^{-4} K^{-1}]	r_m [nm]	$\hat{S}$ [nm^2]	H_{HB}^0 [kJ mol^{-1}]	H_{MF}^0 [kJ mol^{-1}]	exp. T_g	calc. T_g (I)	calc. T_g (II)
[C$_4$C$_1$Py]$^+$	6.58	0.754	4.21	-4.17	28.8	-84.2	-84.7	-83.4
[C$_6$C$_1$Py]$^+$	6.68	0.779	4.61	-4.05	30.7	-82.2	-77.8	-79.4
[C$_8$C$_1$Py]$^+$	6.74	0.801	5.00	-3.89	32.4	-80.2	-72.8	-78.5
[C$_6$Py]$^+$	6.52	0.767	4.41	-5.00	29.6	-77.2	-81.2	-80.2
[C$_6$(C$_1$)$_2$Py]$^+$	6.77	0.791	4.80	-3.19	31.8	-76.2	-74.8	-77.0
[C$_6$C$_1$DmaPy]$^+$	6.82	0.808	5.05	-2.01	33.0	-72.2	-67.3	-73.1
[C$_6$DmaPy]$^+$	6.78	0.804	4.90	-2.47	32.2	-69.2	-70.1	-71.9
[C$_6$C$_3$(C$_2$)$_2$Py]$^+$	6.84	0.839	5.53	-1.70	34.7	-67.2	-65.7	-64.0
[C$_6$C$_2$(C$_1$)$_2$Py]$^+$	6.77	0.810	5.05	-1.72	33.2	-66.2	-71.0	-67.5
[(C$_4$)$_2$Nic]$^+$	6.89	0.808	5.11	-5.34	33.4	-58.2	-68.3	-56.7
[C$_4$Py]$^+$	6.63	0.740	4.01	-5.18	27.4	-76.0^a	-88.6	-87.5
[C$_6$MPipPy]$^+$	6.83	0.831	5.44	-3.06	33.9	-55.2^b	-66.6	-72.8

^{a} Taken from ref. [229]; only observed at rapid cooling. All other experimental T_g were taken from ref. [177]. ^{b} Water content unknown.

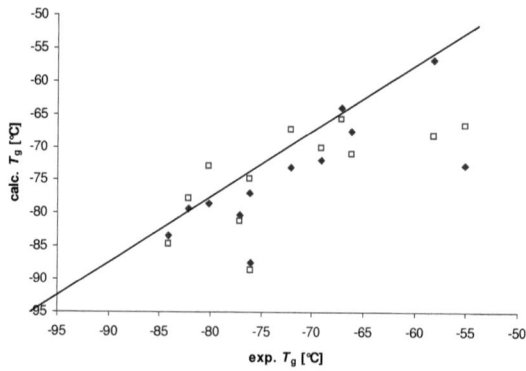

Figure 76: The comparison of experimental and calculated glass transition temperatures. Filled squares belong to data from eqn. (64), hollow ones to data from eqn. (57).

4.9. The Critical Micelle Concentration[267]

4.9.1. Background

Many ionic liquids (ILs) are structurally analogous to surfactants. That is, its anions or cations often consist of a charged hydrophilic head group and a hydrophobic tail domain (see Figure 77 and Figure 78). This characteristic amphiphilic composition of ILs indicates that surface behavior will influence the properties of the systems containing these compounds. The understanding of the molecular interface interactions of ILs in aqueous solutions is a prerequisite for sustainably predicting, controlling and designing IL properties for application in industrial scale processes. Prior to their distribution in the environment, a chemical fate analysis should be performed.[309] Therefore the knowledge of the aggregation behavior of ILs is a vital part of understanding how these compounds participate as components in an aqueous mixed solvent system. A number of researchers have investigated micelle formation in aqueous solutions of ILs.[310,311,312,313,314] To this date, aggregation behavior of imidazolium derivatives with variable chain length and mainly chloride and bromide anions were presented.[310,313,315,316] Several methods to determine the critical micelle concentration (CMC) of ILs in aqueous solution were used including tensiometry, conductometry, small angle neutron scattering, turbidity, potentiometry etc. The results suggest that micelle formation of 1-alkyl-3-methylimidazolium derivatives in water takes place, and that elongation of the alkyl chain in the imidazolium cation decreases the CMC similar to typical cationic surfactants. The CMC will also influence biological processes like biodegradation. The process of micellization can be directly correlated with the interaction of the amphiphiles with non-polar surfaces such as micelles or cell membranes. Since the toxicity of amphiphiles is dominated by their physicochemical properties, and therefore their ability to self-assemble, the CMC provides an important indicator to biodegradation.[317]

Huibers found correlations to predict the CMC of anionic surfactants in water, depending on QSPR studies.[318] Since QSPR methods always rely on a multitude of different descriptors (e.g. chain length or number of functionalized groups) which are restricted to certain substance classes and need to be optimized using training sets from existing measurements, they are not universally applicable. Nagarajan made an extensive first-principles approach to describe the thermodynamics of surfactant aggregation.[319] However, he strongly differentiates between tail and head groups (a division which is not necessarily valid for more complicated compounds) and gives different models for different classes of surfactants. In this project, however, we were interested in making a simple and general approach suited for all types of ionic surfactants.

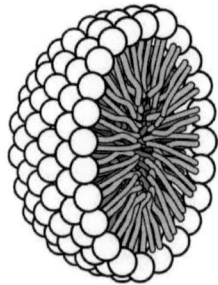

Figure 77: Cross section of a micelle of a molecular surfactant with the polar heads pointing towards the outside; the non-polar tails are concentrated on the inside. Note that micelles don't necessarily have to be spherical.

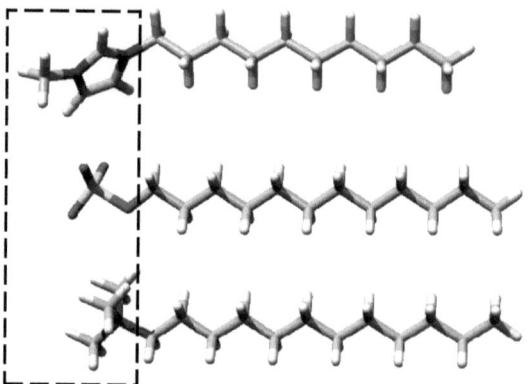

Figure 78: Examples for substituted imidazolium, alkylsulfate, and ammonium ions; shown inside the dashed region are the polar head groups.

4.9.2. Results

Given the typical univalent charges of IL ions, V_m may also be viewed as an indicator for hydrophobicity and thus should also correlate with the solubility of ILs in water and their tendency to aggregate and form micelles. To establish a correlation, we calculated the ionic volumes, V_{ion}^+ and V_{ion}^-, both taken from a BP86/TZVP COSMO calculation, scaled with a linear fit. We chose this method as it is very reliable and there are only very few crystal structures available that contain the investigated cations to infer the molecular volumes from.[320] First, we investigated 30 compounds with micelle-forming cations, i.e. 1-methyl-3-alkyl-imidazolium, alkyl-trimethyl-ammonium and alkyl-pyridinium. In detail, we explored [C_nMIm]$^+$ chlorides and bromides, with n = 6, 8, 10, 12, 14, 16; [C_nMIm]$^+$ bromides, with n = 2, 4, 9; [C_nMIm][BF$_4$], with n = 4, 12; [C_8MIm]I, [C_{18}MIm]Cl; [$N_{1,1,1,n}$]$^+$ chlorides and bromides, with n = 10, 12, 14, 16; [$N_{1,1,1,8}$]Br, [$N_{1,1,1,18}$]Cl; and [C_8Py]Cl. Indeed, we found the CMC to be exponentially dependent on V_m according to:

(65) $\ln(\text{CMC [mmol l}^{-1}\text{]}) = a \cdot V_m \text{ [nm}^3\text{]} + b$

with a = −26.3 and b = 13.7 being empirical constants derived from best fit. The rmse for this fit is 0.339 logarithmic units. Since the correlation is exponential, we took the geometric mean of the known CMC values collected in Table 77 as a basis for our fit. When we further investigated sodium alkyl sulfates (Na[C_nSO$_4$] with n = 8, 10, 12, 14) and double amphiphilic compounds ([C_4MIm][OTf] and [C_4MIm][C_8SO$_4$]), we found them to pose exceptions to the rule. The sodium compounds, however, formed an independent series and correlated nicely with V_m as well (rmse = 0.109), but with different constants a and b (−31.9 and 13.5). Figure 79 is a graphical representation of these findings.

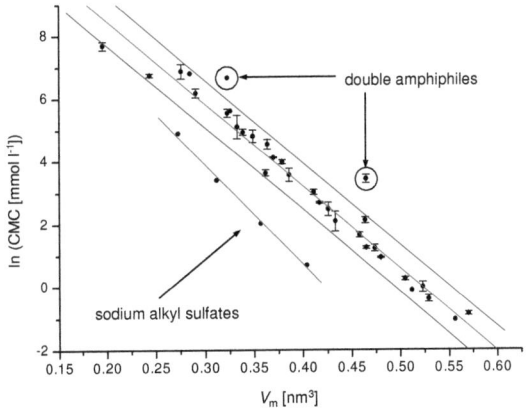

Figure 79: The correspondence of calculated V_m and measured CMC of 30 cationic surfactants (imidazolium-, ammonium- and pyridinium-based ionic liquids; shown along with the 95% prediction bands), four anionic surfactants (sodium alkyl sulfates) and two "mixed" surfactants ([C_4MIm][OTf] and [C_4MIm][C_8SO$_4$]). The latter two do not fit with any of the former series. Error bars are shown where they could be determined.

To eliminate all exceptions, we sought for a better description of the charge density and included the solvent-accessible surface $\hat{S}$. For a combination of $\hat{S}$ and V_m, i.e.

(66) $\ln(\text{CMC [mmol l}^{-1}\text{]}) = a \cdot V_m \text{ [nm}^3\text{]} + b \cdot \hat{S} \text{ [nm}^2\text{]} + c$

we found rmse = 0.479, for the first time including all alkyl sulfates.

Instead of V_m, we then considered the cubed molecular radius, r_m^3, to represent the size of a compound. Then we found the rmse to decrease to 0.325 logarithmic units. This means that the dependency of the CMC on the alkyl chain length is better captured with both r_m^3 and $\hat{S}$, and that the impact of the variation of the head group diminishes.

So we were left with [C$_4$MIm][OTf] and [C$_4$MIm][C$_8$SO$_4$] as problem cases: They are double amphiphilic, meaning both cation and anion contain hydrophobic and hydrophilic groups. At this point, we started to use further ion-specific output from COSMO-RS to model residual interactions. In this case, we included the enthalpies of mixtures calculated by COSMO-RS in the pure liquid state (enthalpies in a 1:1 mixture of cation and anion as well as both cation and anion in infinite dilution in water). All enthalpies were calculated at 25 °C.

Overall and converged, we found the CMC to be depending on the following expression:

$$(67) \quad \ln(CMC) = a \cdot r_m^3 + b \cdot \hat{S} + \sum_i \frac{c_i H_i^{diff}}{\hat{S}} + d \cdot H_{ring} + e$$

with a...e being empirical constants derived from best fit. The coefficients are given in Table 75; all descriptors (r_m, $\hat{S}$, H_i^{diff}, H_{ring}) are included in Table 76. With this equation, the remaining two exceptions were eliminated and the overall error greatly reduced. Here, rmse = 0.191 logarithmic units; $r^2 = 0.994$.

Table 75: The coefficients for eqn. (67) and their values of best fit.

coefficient	a	b	c_i			d	e
unit	nm^{-3}	nm^{-2}	nm^2 mol kJ^{-1}			mol kJ^{-1}	
i			vdW	MF	HB		
value	17.48	−4.44	0.29	−0.17	0.07	−0.09	15.36

Table 76: Molecular volumes and radii as well as the solvent-accessible surface and different enthalpies (all in kJ mol^{-1}) calculated with COSMO-RS.

	V_m [nm^3]	r_m [nm]	$\hat{S}$ [nm^2]	H_{MF}^{diff}	H_{vdW}^{diff}	H_{HB}^{diff}	H_{ring}
[N$_{1,1,1,8}$]Br	0.323	0.628	3.21	25.0	−13.1	97.2	0.00
[N$_{1,1,1,10}$]Cl	0.364	0.640	3.55	25.3	−11.3	111.1	0.00
[N$_{1,1,1,10}$]Br	0.370	0.650	3.60	22.3	−14.2	97.4	0.00
[N$_{1,1,1,12}$]Cl	0.411	0.659	3.95	22.7	−12.3	111.5	0.00
[N$_{1,1,1,12}$]Br	0.417	0.670	4.01	20.0	−15.3	97.8	0.00
[N$_{1,1,1,14}$]Cl	0.458	0.678	4.34	20.1	−13.3	111.6	0.00
[N$_{1,1,1,14}$]Br	0.465	0.688	4.40	17.7	−16.3	98.0	0.00
[N$_{1,1,1,16}$]Cl	0.505	0.694	4.74	17.6	−14.4	111.6	0.00
[N$_{1,1,1,16}$]Br	0.512	0.705	4.80	15.5	−17.3	98.0	0.00
[N$_{1,1,1,18}$]Cl	0.556	0.711	5.15	15.4	−15.4	112.1	0.00
[C$_2$MIm]Br	0.196	0.553	2.21	32.5	−13.0	89.2	−4.09
[C$_4$MIm]Br	0.244	0.585	2.61	27.5	−13.8	89.8	−4.09
[C$_4$MIm][BF$_4$]	0.276	0.627	2.93	9.1	−2.6	31.8	−4.09

[C$_4$MIm][OTf]	0.323	0.672	3.30	7.7	-3.5	47.4	-4.09
[C$_4$MIm][C$_8$SO$_4$]	0.465	0.761	4.60	10.0	-10.2	72.3	-4.09
[C$_6$MIm]Cl	0.285	0.601	2.95	26.6	-9.8	101.2	-4.09
[C$_6$MIm]Br	0.291	0.612	3.00	23.6	-14.8	90.0	-4.09
[C$_8$MIm]Cl	0.333	0.625	3.34	23.5	-10.7	101.6	-4.09
[C$_8$MIm]Br	0.339	0.636	3.40	20.8	-15.7	90.5	-4.09
[C$_8$MIm]I	0.349	0.651	3.49	15.2	-15.6	71.1	-4.09
[C$_9$MIm]Br	0.362	0.646	3.60	19.4	-16.2	90.5	-4.09
[C$_{10}$MIm]Cl	0.379	0.646	3.74	20.8	-11.7	102.2	-4.09
[C$_{10}$MIm]Br	0.386	0.657	3.80	18.4	-16.7	91.0	-4.09
[C$_{12}$MIm]Cl	0.426	0.665	4.14	18.2	-12.6	102.3	-4.09
[C$_{12}$MIm]Br	0.433	0.676	4.20	15.9	-17.7	91.2	-4.09
[C$_{12}$MIm][BF$_4$]	0.464	0.717	4.52	4.8	-6.0	32.7	-4.09
[C$_{14}$MIm]Cl	0.474	0.683	4.54	15.5	-13.6	102.4	-4.09
[C$_{14}$MIm]Br	0.480	0.694	4.60	13.5	-18.7	91.3	-4.09
[C$_{16}$MIm]Cl	0.523	0.700	4.94	13.0	-14.7	102.5	-4.09
[C$_{16}$MIm]Br	0.529	0.711	5.00	11.2	-19.6	91.4	-4.09
[C$_{18}$MIm]Cl	0.570	0.716	5.33	10.6	-15.7	102.7	-4.09
[C$_8$Py]Cl	0.326	0.622	3.27	24.7	-10.8	106.4	-4.90
Na[C$_8$SO$_4$]	0.273	0.507	2.96	19.9	-5.0	83.9	0.00
Na[C$_{10}$SO$_4$]	0.312	0.525	3.37	17.5	-6.1	84.0	0.00
Na[C$_{12}$SO$_4$]	0.357	0.545	3.76	14.9	-7.0	84.1	0.00
Na[C$_{14}$SO$_4$]	0.404	0.564	4.16	12.2	-8.1	84.2	0.00

The sum of the c_i H_i^{diff} denotes the difference of the interaction of the surfactant's ions with water to the interaction of both ions in the pure compound (which is idealized as being liquid: a valid assumption, since, at the CMC, aggregates possess no long range order). The different enthalpy contributions that COSMO-RS calculates are denoted with index i; they are hydrogen bonding (HB), van-der-Waals (vdW), and misfit (MF) interaction. All of these enthalpies are scaled by the reciprocal surface area. A correction for ring size, H_{ring}, also put out by COSMO-RS, is included with the coefficient d; this term is independent from solvent and temperature.

The terms in eqn. (67) are highly independent. The strongest correlation by far is between $H_{MF}^{diff}/\hat{S}$ and $H_{HB}^{diff}/\hat{S}$, with r^2 = 0.830, however the physical realities behind these terms are not interconnected. The second-strongest is between $H_{MF}^{diff}/\hat{S}$ and $\hat{S}$, with r^2 = 0.628. As all other internal correlations are between 0 and 0.5, we believe that none of the terms could be omitted.

Table 77 and Figure 80 show the quality of the predictions. Since the correlation is exponential, errors are quite large for higher CMC values; however, the error bars are in the same order of magnitude as the errors of the experimental CMC determination.

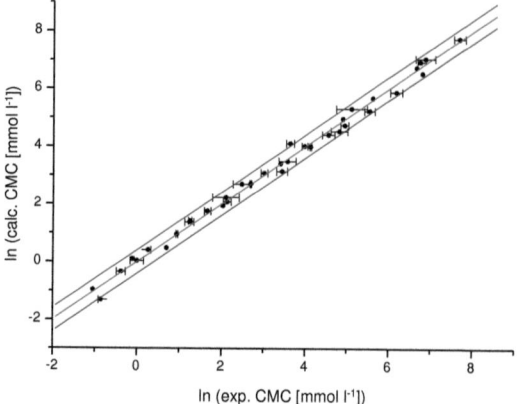

Figure 80: The correspondence of the experimental and calculated CMC of all 36 surfactants tested, shown along with error bars and the 95% prediction bands. Calculations were performed using eqn. (67).

Table 77: The independently experimentally measured versus the extrapolated CMCs [mmol l^{-1}] from eqn. (67) of different ILs measured at room temperature.

compound	surface tension	conductivity	other	extrap.
[N$_{1,1,1,8}$]Br			225[321]	188
			290[322]	
[N$_{1,1,1,10}$]Cl	70[312]	94.7[323]		82
[N$_{1,1,1,10}$]Br		62.7[324]	62[321]	55
			60.2[324]	
[N$_{1,1,1,12}$]Cl	18[312]	22.2[323]		22
		21.3[325]		
[N$_{1,1,1,12}$]Br			14.3[321]	15
			15[326]	
[N$_{1,1,1,14}$]Cl	5.5[327]	5.63[323]		5.8
	4.5[312]	5.5[328]		
[N$_{1,1,1,14}$]Br		3.8[328]	3.5[326]	4.1
[N$_{1,1,1,16}$]Cl	1.3[312]	1.46[312]		1.5
[N$_{1,1,1,16}$]Br			0.9[326]	1.1
[N$_{1,1,1,18}$]Cl			0.35[329]	0.38
[C$_2$MIm]Br	2500[330]	1900[330]		2316
[C$_4$MIm]Br	800[330]	900[330]		1025
[C$_4$MIm][BF$_4$]	800[313]	820[313]		1153
	1370^a			
[C$_4$MIm][OTf]	782^a			843
[C$_4$MIm][C$_8$SO$_4$]	40.5^a	31[311]		23
[C$_6$MIm]Cl	900[312]			689
[C$_6$MIm]Br	600[330]	400[330]	880[331]	361
	470[316]		800[331]	
[C$_8$MIm]Cl	220[310]	234[310]	200[312]	203
	100[313]	90[313]		
	220[312]			
[C$_8$MIm]Br	150[330]	150[330]	180[331]	114
	121[316]		190[331]	
[C$_8$MIm]I	100[313]	150[313]		93

[C$_9$MIm]Br	40[330]	30[330]		61
		74[314]		
[C$_{10}$MIm]Cl	59.9[310]	53.8[310]	45[312]	56
	55[312]	40.47[332]	55[312]	
	39.9[332]			
[C$_{10}$MIm]Br	20[316]	40[315]	42[331]	33
	29.3[333]	41[314]	46[331]	
		32.9[333]		
[C$_{12}$MIm]Cl	15[312]	13.47[332]	7[312]	15
	13.17[332]		13[312]	
[C$_{12}$MIm]Br	4.3[316]	9.8[314]	10[331]	9.3
		8.5[333]	12[331]	
		9.5[334]		
[C$_{12}$MIm][BF$_4$]	9.2[333]	7.6[333]		8.0
[C$_{14}$MIm]Cl	4[312]	3.15[310]	3[312]	4.0
	3.4[310]	3.68[332]	4[312]	
	2.98[332]			
[C$_{14}$MIm]Br		2.5[314]		2.6
		2.6[334]		
[C$_{16}$MIm]Cl	1.3[310]	1.14[310]		1.0
	0.88[335]	0.86[332]		
	0.87[332]			
[C$_{16}$MIm]Br		0.8[316]		0.71
		0.61[314]		
		0.65[334]		
[C$_{18}$MIm]Cl	0.4[310]	0.45[310]		0.27
[C$_8$Py]Cl		274[310]		292
Na[C$_8$SO$_4$]			134[336]	143
Na[C$_{10}$SO$_4$]			30[336]	30
Na[C$_{12}$SO$_4$]			7.6[337]	7.0
Na[C$_{14}$SO$_4$]			2.00[336]	1.6

[a] Measured by J. Łuczak in cooperation with the Chemical Faculty of the Gdańsk University of Technology, Poland.

At this point, we noted a striking similarity to the thermodynamics of crystal formation. In a solution, the Gibbs energy of crystal formation ($\Delta_c G$) is composed of the Gibbs energies of phase transition ($\Delta_n G$) and phase boundary formation ($\Delta_\gamma G$), augmented by an – often disregarded – elasticity term ($\Delta_e G$) which models the influence of adjacent particles within a rigid phase. This relation has entered into many textbooks and can in principle be traced back to the studies of Gibbs in the 19th century[267,267]:

(68) $\Delta_k G = \Delta_n G + \Delta_\gamma G + \Delta_e G$

$\Delta_n G$ is proportional to ar^3, where r is the radius of the crystal nucleus, $a < 0$ (chemical interactions are formed in the three dimensions of a crystal, which is energetically advantageous); $\Delta_\gamma G$ is proportional to br^2, with $b > 0$ (a new surface has to be formed at the phase boundary, which is energetically disfavored). So we get:

(69) $\Delta_k G = ar^3 + br^2 + \Delta_e G$

For the Gibbs energy of micelle formation, we can set:

(70) $\Delta_m G = -RT \ln K$

where $K = \ln(a(\text{micellized surfactant})/a(\text{solvated surfactant}))$ and a = activity. Assuming T is constant and the concentration of the solvated surfactant is in high excess, making it quasi-constant, we obtain:

(71) $-\Delta_m G \propto \ln(a(\text{micellized surfactant}))$

At the critical point, we can substitute $\ln(\text{CMC})$ in eqn. (67) for $-\Delta_m G$ and it becomes clear that $\Delta_k G$ and $\Delta_m G$ have the same constituents if r^3 in eqn. (69) gets substituted for r_m^3, r^2 for $\hat{S}$, and the different interaction enthalpies take the part of the elastic contributions $\Delta_e G$. Also, as expected from eqn. (71), the signs of a and b in eqn. (67) and (69) are opposite. Therefore, crystal and micelle formation seem to follow similar rules – a finding that deserves its own further investigation.

In eqn. (67), the coefficients c_i may include repulsive and attractive forces, and may include contributions that stem from overcompensation of the other coefficients, especially a and b. Also, the interaction enthalpies are not physically verifiable and therefore, their size may not mirror the real nature of the interaction. Combined, however, they are well suited to describe the missing parts of the correlation.

For data collected between 23 and 25 °C, as well as data with no given temperature, standard conditions were assumed. We did not take the study of Modaressi et al. into account since their measurements, especially for [C_{10}MIm]Cl and [C_{12}MIm]Br, deviate strongly from those of all other authors. Their methodology was unclear, and the determination of the breakpoint the authors describe is questionable.[338]

There are several sources of experimental uncertainties which can explain the large variation in some of the measured data. The typical CMC measurements present not a single value, but rather a range during which the phenomenon of micelle formation takes place. This generates some difficulty in comparing results measured with different methods. Also, traces of organic compounds (grease, 1-methylimidazol) can lead to a decline in surface tension, thus increasing the CMC; traces of inorganic salts (halides) can influence the ion product of water, therefore decreasing the CMC. Both imperfections can be introduced by the typical production processes of ionic liquids.

4.9.3. Conclusion

We found a relation for the prediction of the aqueous critical micelle concentration of cationic, anionic and double amphiphilic surfactants. For the 30 cationic surfactants tested – comprising alkyl-substituted ammonium and imidazolium compounds and one pyridinium compound –, a simple linear dependence on the molecular volume, V_m, was found to sufficiently describe the CMC. This is in good agreement with the Stauff-Klevens rule that correlates the CMC with the chain length. To also include anionic and double amphiphilic surfactants, a linear extension of the equation with additional descriptors (molecular surface area, mixing enthalpies) was necessary. To determine these descriptors, we employed COSMO-RS and could successfully predict the CMC of 36 compounds where it does not matter whether the micelles are formed by cations or anions or both. The resulting formula resembles the Gibbs relation for crystal formation. In contrast to QSPR methods, no individual training sets are needed for our method, much broadening the applicable range.

4.10. Further Considerations about IL Nanostructures

Water-free ILs with longer alkyl chains seem to form some kind of "foam", where the non-polar alkyl chains form the surface and the polar parts, including the anion, occupy the interstices.[103,104] Even the formation of proper micelles was predicted (via MD simulations) in the bulk of ILs with long side chains (see Figure 81).[339] However, these micelles are not sharply bordered and blend with their surroundings, so the experimental evidence is hard, if not impossible, to find. There could be an indirect proof of this nanostructuring, however: Due to the pre-ordering of long-chain ILs in micelle-like shapes, less reorientation would have to occur if a polar molecule like water was added. Therefore, for example the melting point would not change as dramatically as for short-chained ILs.

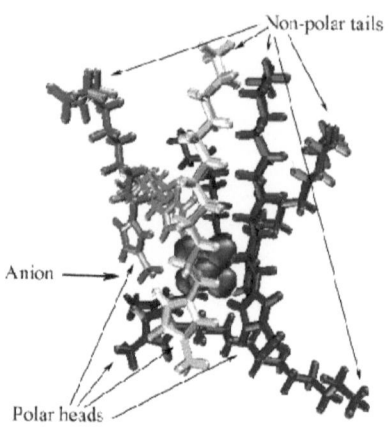

Figure 81: A micelle-like structure of [C$_{12}$MIm][PF$_6$] as predicted from MD simulation. Picture taken from ref. [339]. The anion forms the center of a cage built from the cations.

Deduced from measurements of capacitance and adsorption isotherms, micelle-like structures (hemicelles and admicelles) were also predicted on the surfaces of charged electrodes and minerals (see Figure 82).[340,341] An experimental study of the alkyl chain tilt angles of imidazolium ILs absorbed on SiO$_2$ seems to underline this theory.[342]

The interaction of ionic liquids with a porous medium like soil is usually by Coulomb interaction of the charged head group with the negative charge of the clay surface, or with the organic fraction of the porous media and lipophilic tail of the IL.[340] Transport parameters such as the sorption coefficient (K_d) form an important part to model the spreading of chemicals in the environment and through the solid porous phase.[343] Therefore, the determination of these parameters is essential. However, the level of knowledge of the transport parameters of ionic liquids in porous media is yet limited.

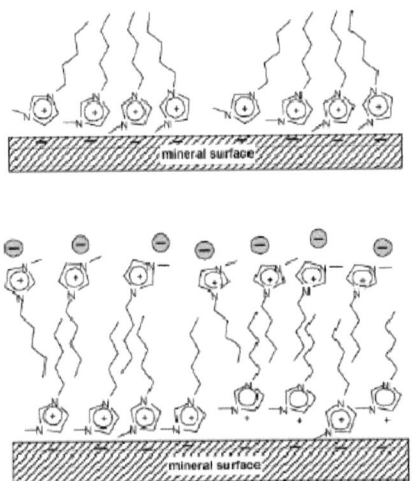

Figure 82: Self-ordering on surfaces, as deduced from adsorption isotherms. Picture taken from ref. [340].

We found in cooperation with C. Jungnickel et al. that K_d as the ratio of adsorbed to dissolved surfactant at the equilibrium point depends on the molecular volume.[343] The chlorides of $[C_nMIm]^+$ (n = 2...6 and 8), $[C_4C_1MIm]^+$ and $[C_4C_1Py]^+$ on 21 very different soil types were investigated. The resulting formula for the calculated K_d is:

$$(72) \quad K_d = \frac{a \cdot \text{CEC} + b \cdot \text{OC} \cdot R_T}{V_m}$$

with r² = 0.8985, where CEC = cation exchange capacity, OC = organic carbon content, R_T = retention time. These factors were gathered from routine measurements to characterize soil types. The retention time is a measure for lipophilicity and is determined chromatographically. The organic carbon content can be determined by combustion analysis. The CEC is a measure of how many cations a particular type of soil can replace and is determined by extracting all cations with a neutral salt. Once again, V_m was calculated with BP86/TZVP+COSMO.

5. Experimental part

5.1. Experimental Techniques

5.1.1. General Procedures and Starting Materials

Due to air- and moisture sensitivity of most materials all manipulations (if not mentioned otherwise) were undertaken using standard vacuum and Schlenk techniques as well as a glove box with an argon or nitrogen atmosphere (H_2O and O_2 < 1 ppm). All solvents were dried by conventional drying agents and distilled afterwards.

5.1.2. Analytical Methods

NMR spectra were collected on a Bruker Avance II WB 400 MHz or a Bruker Avance DPX 200 MHz. Where applicable, toluene-d8 was added in a sealed lock capillary. IR spectra were recorded on a Nicolet Magna-IR 760 spectrometer using a diamond Orbit ATR unit (extended ATR correction with refraction index 1.5 was used). Raman spectra were collected at RT on a Bruker RAM II FT-Raman spectrometer (using a liquid nitrogen cooled Ge detector) in sealed melting point capillaries.

Melting points and enthalpies were determined with a Setaram DSC 131 in 30 µl aluminium crucibles with an empty crucible of the same type as reference. The crucibles were filled and closed in a glovebox and then stored under Argon. For processing, the software SETSOFT 2000 version 3.0.2 was used. The melting points were determined from the onset points of the melting process. Melting enthalpies were calculated by integrating the melting peaks.

[C_2MIm][NO_3] (98%) and [C_8MIm][Dca] (99%) were bought from iolitec and dried while stirring at 70 °C for 3 d in a dynamic vacuum (< 100 µbar) before determining the melting point and enthalpy.

Data collection for X-ray structure determinations were performed on a BRUKER SMART diffractometer equipped with an APEX II detector using graphite-monochromated Mo-K$_\alpha$ (0.71073 Å) radiation. Single crystals were mounted in perfluoroether oil on top of a glass fiber and then brought into the cold stream of a low temperature device so that the oil solidified. All calculations were performed using the SHELX97 software package. The structures were solved by the Patterson heavy atom method or direct methods and successive interpretation of the difference Fourier maps, followed by least-squares refinement. All non-hydrogen atoms were refined anisotropically. The hydrogen atoms were included in the refinement in calculated positions by a riding model using fixed isotropic parameters.

ESI mass spectra were collected with a LCQ-Advantage spectrometer by Thermo.

5.2. Synthesis and Characterization of the Prepared Compounds

5.2.1. Synthesis of Li[Nb(hfip)$_6$]

Li-hfip was synthesized according to a known procedure from LiH and H-hfip.[344] In modification of the published procedure,[345] 5.907 g (21.86 mmol) freshly sublimed yellow NbCl$_5$ (Aldrich, 99%) was suspended in 400 ml 1,2-dichloroethane in a 1 l round bottom flask fitted with a reflux condenser (cooled to 0 °C). 22.824 g (131.2 mmol) Li-hfip were added. After 1 h of vigorous stirring at room temperature, the mixture was heated to reflux and kept refluxing for 4 h. After cooling to RT, the mixture was filtered. The insoluble substance was washed three times with 100 ml 1,2-dichloroethane and one time with 200 ml perfluorohexane. From the combined filtrates, the remaining solvent was removed in vacuo yielding 15.724 g (65.3%) pure, off-white product.

^{1}H NMR (200.13 MHz, 1,2-C$_6$H$_4$F$_2$/ toluene-d8): 4.74 (m, br); ^{19}F NMR (188.31 MHz, 1,2-C$_6$H$_4$F$_2$/ toluene-d8): -75.54 (s); ^{7}Li NMR (155.52 MHz, 1,2-C$_6$H$_4$F$_2$/ toluene-d8): -0.75 (s); ^{93}Nb NMR (97.95 MHz, 1,2-C$_6$H$_4$F$_2$/ toluene-d8): -1390.5 (s).

IR: $\tilde{\nu}$ = 520.8 (s, δ_{C-F}), 685.5 (s, γ_{C-H}), 700.7 (s, ν_{C-C}), 747.4 (s, ν_{C-C}), 850.7 (s, ν_{C-C}), 891.3 (s, δ_{C-H}), 1097.0 (s, δ_{C-H}), 1120.4 (s, ν_{C-O}), 1189.3(s, ν_{C-O}), 1212.7 (s, γ_{C-H}), 1257.5 (s, δ_{C-H}), 1285.1 (m, γ_{C-H}), 1364.7 (m, γ_{C-H}), 2953.4 (vw, ν_{C-H}) cm^{-1}

Raman: $\tilde{\nu}$ = 295.81 (δ_{C-O}), 330.97 (δ_{C-O}), 537.66 (δ_{C-F}), 710.47 (ν_{C-F}), 750.40 (ν_{C-C}), 854.79 (ν_{C-C}), 1105.12 (ν_{C-F}), 1138.98 (ν_{C-F}), 1301.51 (γ_{C-H}), 1371.49 (ν_{C-H}), 2947.48 (ν_{C-H}) cm^{-1}.

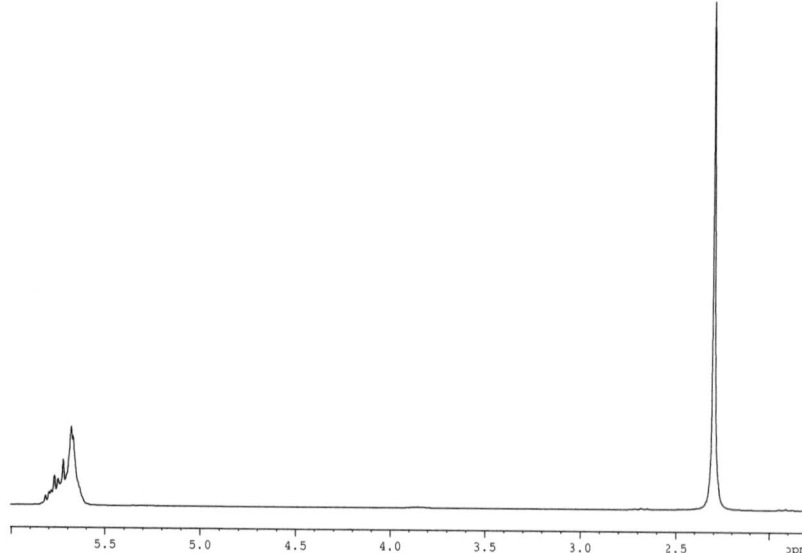

Figure 83: ^{1}H NMR of Li[Nb(hfip)$_6$] in 1,2-C$_6$H$_4$F$_2$/ toluene-d8.

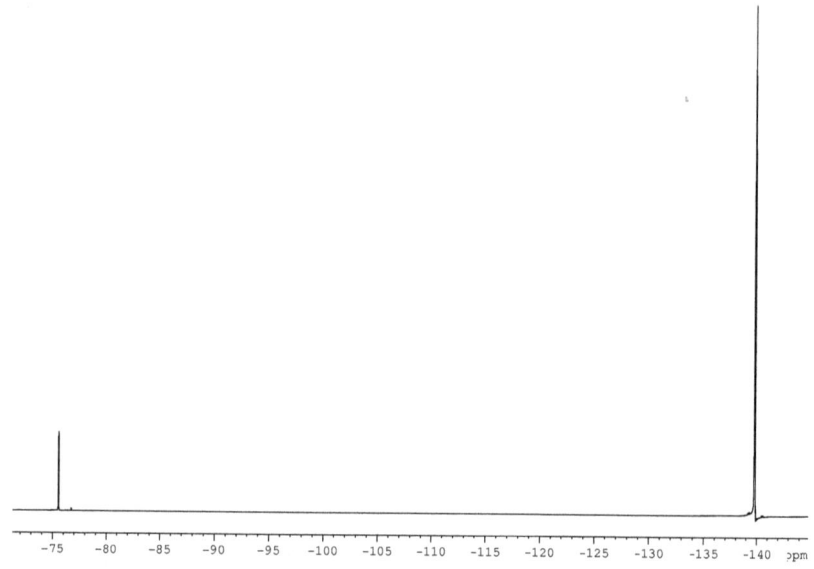

Figure 84: ^{19}F NMR of Li[Nb(hfip)$_6$] in 1,2-C$_6$H$_4$F$_2$/ toluene-d8.

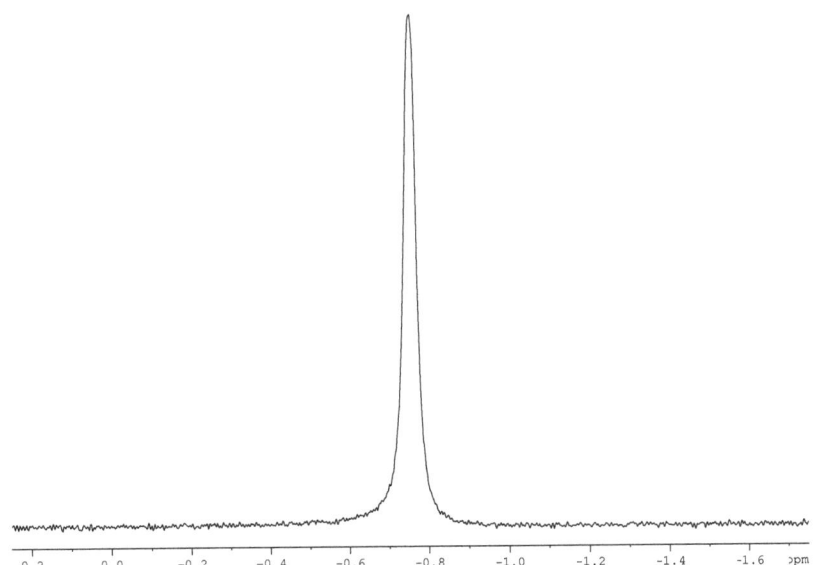

Figure 85: ^{7}Li NMR of Li[Nb(hfip)$_6$] in 1,2-C$_6$H$_4$F$_2$/ toluene-d8.

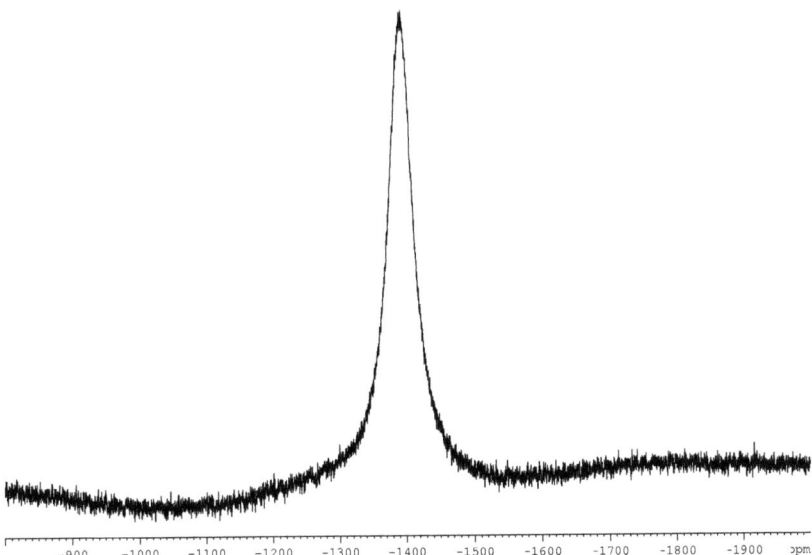

Figure 86: ^{93}Nb NMR of Li[Nb(hfip)$_6$] in 1,2-C$_6$H$_4$F$_2$/ toluene-d8.

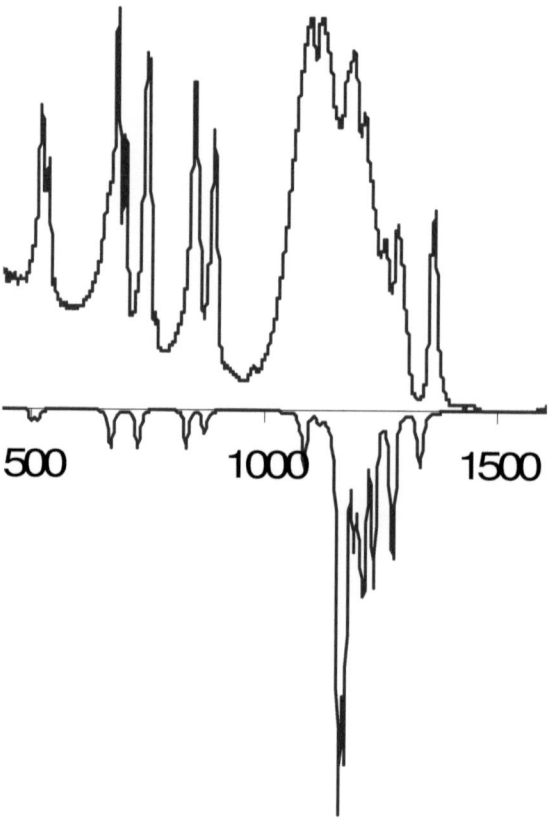

Figure 87: The IR absorption spectra of Li[Nb(hfip)$_6$]. Top, mirrored: measured compound; bottom: calculated anion (BP86/SV(P)); both were scaled to the same maximum intensity.

5.2.2. Synthesis of Li[Ta(hfip)$_6$]

2.504 g (6.990 mmol) freshly sublimed TaCl$_5$ (Acros, 99.85%) was suspended in 150 ml 1,2-dichloroethane in a 250 ml round bottom flask fitted with a reflux condenser (cooled to 0 °C). After 20 min stirring, 7.298 g (41.95 mmol) Li-hfip was added. The mixture was then refluxed for 35 h. After cooling to RT, the suspension was filtered solvent was removed from the filtrate. The raw, colorless product was washed with 5 ml 1,2-dichloroethane and recrystallized from 55 ml of fluorobenzene. Yield 6.4048 g (77.0%) pure product.

^{1}H NMR (200.13 MHz, 1,2-C$_6$H$_4$F$_2$/ toluene-d8): 4.86 (br); ^{19}F NMR (188.31 MHz; 1,2-C$_6$H$_4$F$_2$/ toluene-d8): -75.54 (s); ^{7}Li NMR (155.52 MHz, C$_6$H$_5$F/ toluene-d8): -0.54 (s)

IR: $\tilde{\nu}$ = 519.9 (s, δ_{C-F}), 684.5 (s, γ_{C-H}), 747.8 (s, ν_{C-C}), 849.6 (s, ν_{C-C}), 890.3 (s, δ_{C-H}), 1096.8 (s, δ_{C-H}), 1118.4 (s, ν_{C-O}), 1184.3 (s, ν_{C-O}), 1212.4 (s, γ_{C-H}), 1256.8 (s, δ_{C-H}), 1284.0 (m, γ_{C-H}), 1364.8 (m, γ_{C-H}), 2955.4 (vw, ν_{C-H}) cm^{-1}

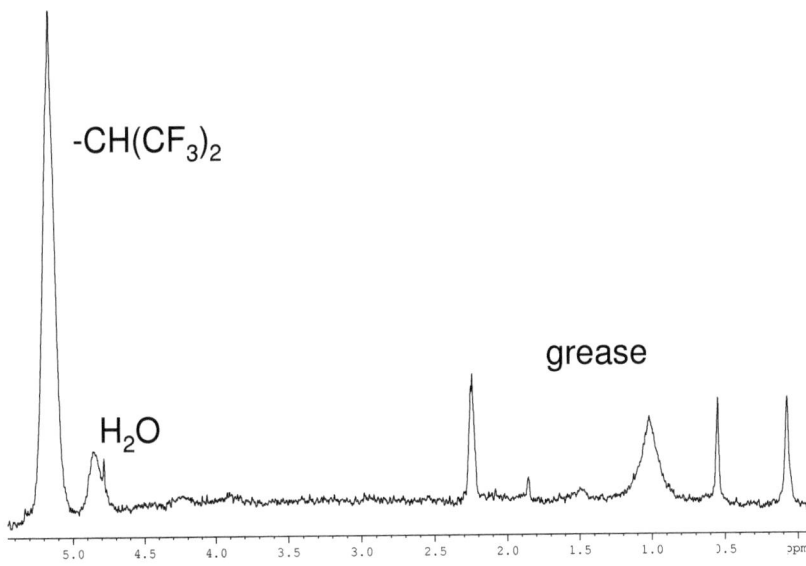

Figure 88: ^{1}H NMR of Li[Ta(hfip)$_6$] in 1,2-C$_6$H$_4$F$_2$/ toluene-d8. Due to the low solubility of the product, the solvent signal was too strong in comparison and therefore left out. N.b. all of the impurities at the right end, including water, are also found in a solvent sample of 1,2-C$_6$H$_4$F$_2$/ toluene-d8 and thus are most likely not contained in the product itself.

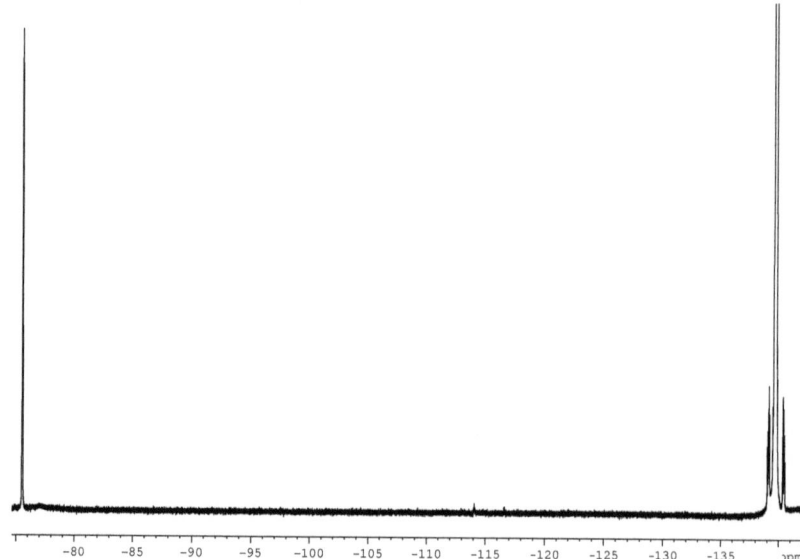

Figure 89: ^{19}F NMR of Li[Ta(hfip)$_6$] in 1,2-C$_6$H$_4$F$_2$/ toluene-d8.

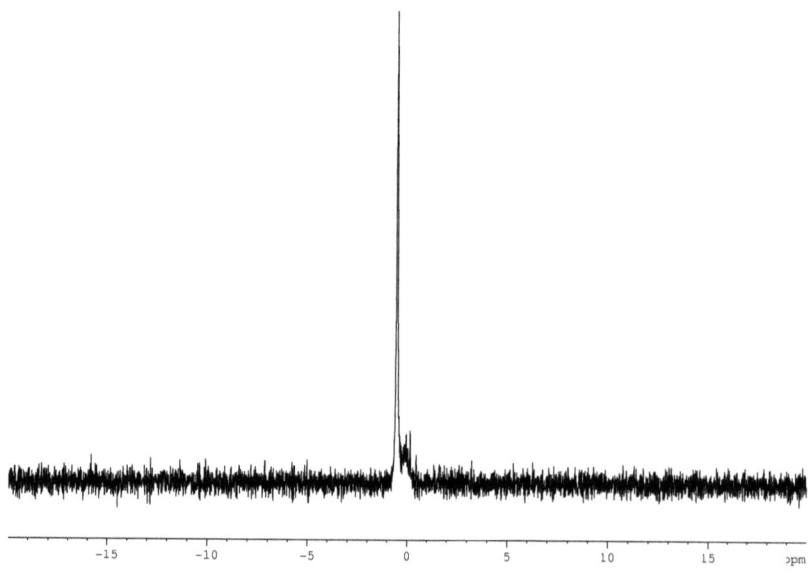

Figure 90: ^{7}Li NMR of Li[Ta(hfip)$_6$] in C$_6$H$_5$F/ toluene-d8.

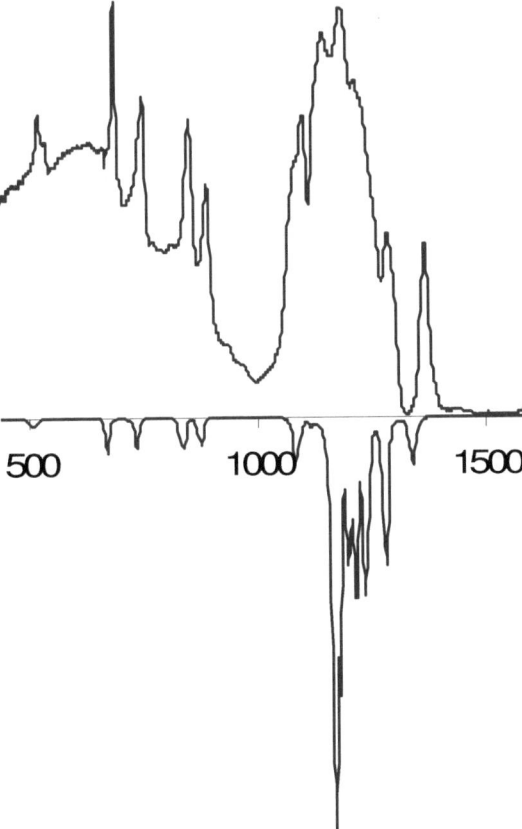

Figure 91: The IR absorption spectra of Li[Ta(hfip)$_6$]. Top, mirrored: measured compound; bottom: calculated anion (BP86/SV(P)); both were scaled to the same maximum intensity.

5.2.3. Synthesis of Ag[Nb(hfip)$_6$]

In a small Young valve Schlenk tube, 56 mg (0.44 mmol) AgF (99%, Apollo Scientific) and 337 mg (0.306 mmol) Li[Nb(hfip)$_6$] were combined in a glove box under argon with the exclusion of light. 4 ml CH$_2$Cl$_2$ were added; the suspension was degassed three times, then sonicated for 3 d in a powerful ultrasonic bath (Bandelin Sonorex RK 514). The suspension was filtered and all volatiles of the soluble filtrate removed in vacuo. Yield 0.3598 g (97.8%) of the pure, light brown, very light-sensitive product.

^{1}H NMR (200.13 MHz, CDCl$_3$): 5.07 (br); ^{19}F NMR (188.31 MHz, CDCl$_3$): -74.55 (s)

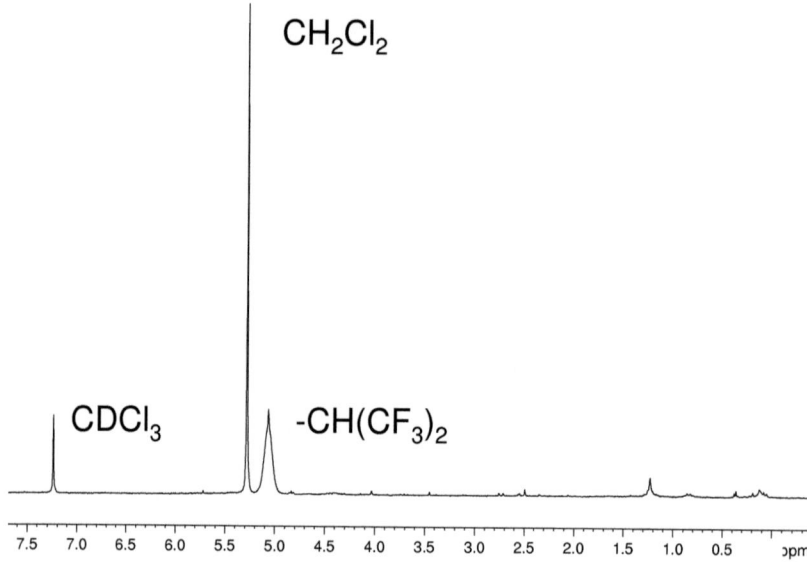

Figure 92: ^{1}H NMR of Ag[Nb(hfip)$_6$] in CDCl$_3$.

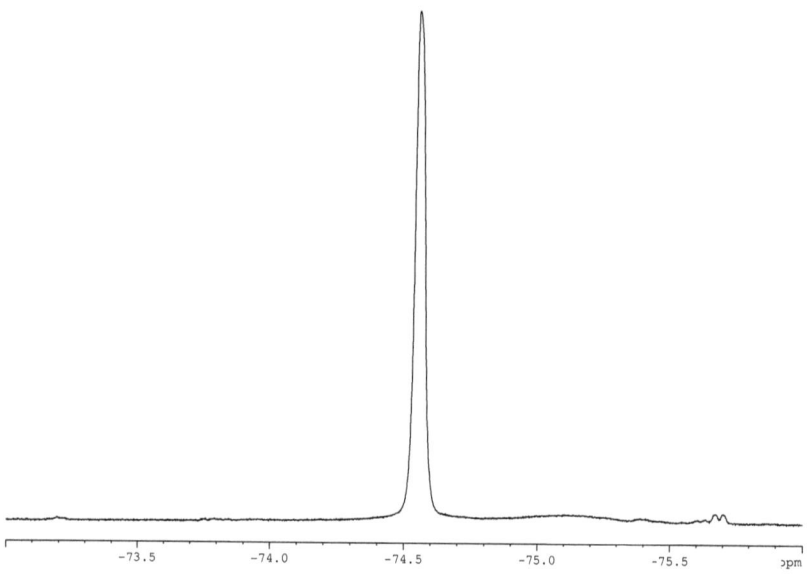

Figure 93: ^{19}F NMR of Ag[Nb(hfip)$_6$] in CDCl$_3$.

IR: $\tilde{\nu}$ = 520.6 (s, δ_{C-F}), 684.5 (s, γ_{C-H}), 747.8 (s, ν_{C-C}), 849.6 (s, ν_{C-C}), 890.3 (s, δ_{C-H}), 1096.8 (s, δ_{C-H}), 1118.4 (s, ν_{C-O}), 1184.3(s, ν_{C-O}), 1212.4 (s, γ_{C-H}), 1256.8 (s, δ_{C-H}), 1284.0 (m, γ_{C-H}), 1364.8 (m, γ_{C-H}), 2954.5 (vw, ν_{C-H}) cm^{-1}

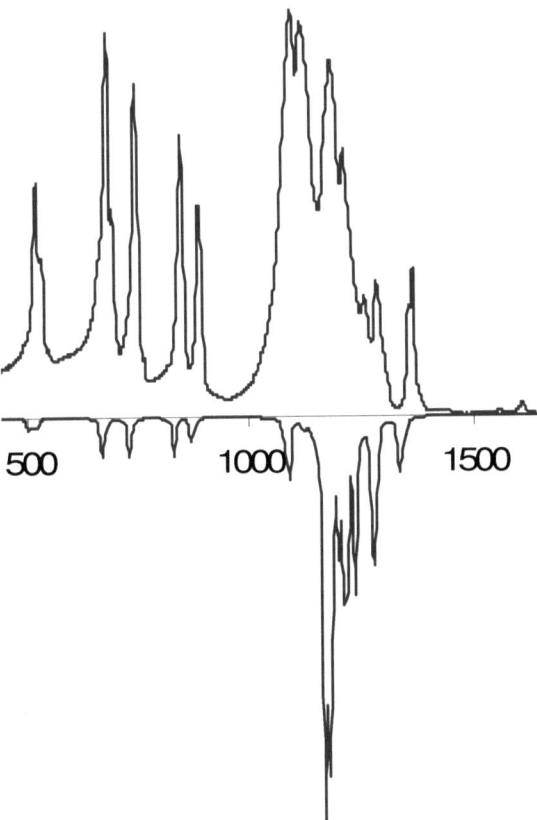

Figure 94: The IR absorption spectra of Ag[Nb(hfip)$_6$]. Top, mirrored: measured compound; bottom: calculated anion (BP86/SV(P)); both were scaled to the same maximum intensity.

5.2.4. Synthesis of Na-pfad

H-pfad reacts with NaOH in aqueous solution;[346] however, the product was not characterized until now. In our preparation, 8.69985 g (20.6 mmol) of H-pfad were dissolved in a mixture of 175 ml methanol and 175 ml distilled water. Whilst stirring, 1.6633 g (41.6 mmol) NaOH, dissolved in 400 ml water, were added over the course of 30 min. The product was extracted six times with 400 ml ether. After evaporation of the unified organic phases, the resulting glassy solid was recrystallized from 100 ml of a 1:1 mixture of ether and light petrol ether. Yield 8.64446 g (94.0%) of the pure, white, solid product which owns a distinct, cinnamon-like odor. Due to its insensitive nature, the whole preparation could take place on air using regular, non-distilled solvents.

In the ^{19}F NMR spectrum of the free alcohol H-pfad, dissolved in CDCl$_3$, there's a singlet at 120.76 ppm (6 CF$_2$) and another one at 221.99 ppm (3 CF).[347] Probably, the additional splitting observed in Na-pfad stems from the fact that MTBE is much stronger coordinated to the Na atom than CDCl$_3$ to H-pfad.

^{19}F NMR (100.62 MHz, MTBE/ toluene-d8): -121.22 (s, 6 F), -123,85 (s, 6 F), -222.63 (s, 3 F)

IR: $\tilde{\nu}$ = 647.8 (m, δ_{C-F}), 952.1 (s, ν_{C-F}), 971.8 (m, δ_{C-C}), 1219.0 (w, ν_{C-F}), 1250.8 (m, ν_{C-F}), 1267.6 (s, τ_{C-C}) cm^{-1}

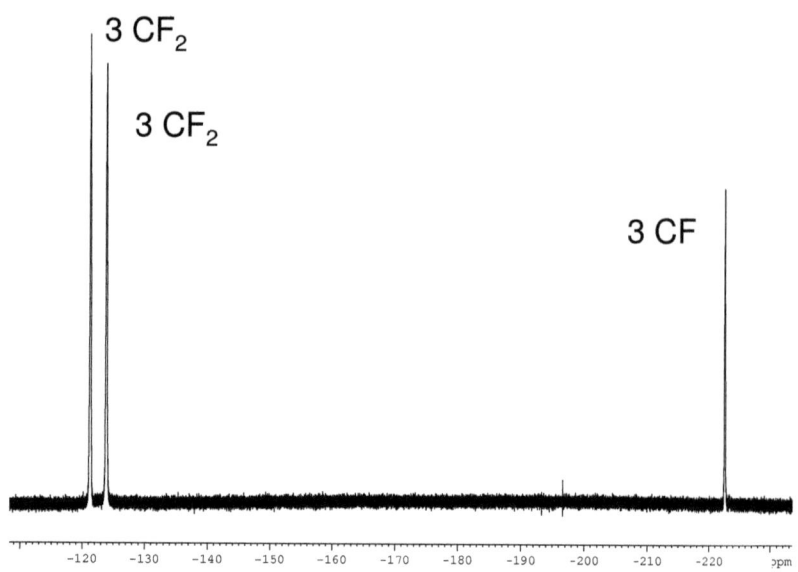

Figure 95: ^{19}F NMR of Na-pfad in MTBE/ toluene-d8.

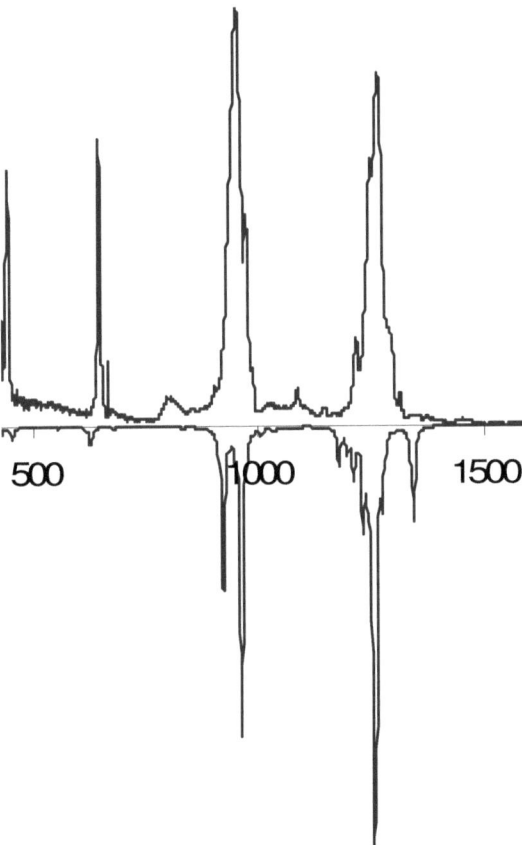

Figure 96: The IR absorption spectra of Na-pfad. Top, mirrored: measured, bottom: calculated (BP86/SV(P)) compound; both were scaled to the same maximum intensity.

5.2.5. Synthesis of Na[Al(pfad)$_4$]

2.024 g (4.558 mmol) of Na-pfad was added to 0.153 g AlCl$_3$ (1.15 mmol; doubly sublimed) in a 500 ml round bottom flask fitted with a reflux condenser (cooled to 0 °C). The mixture was suspended in 250 ml 1,2-dichloroethane and refluxed for 6 d. After cooling to RT, the solvent was removed by filtration and the raw, white product washed with 200 ml 1,2-dichloroethane and then extracted with MTBE; yield 0.2084 g (10.5%) of a white solid. This reaction was not yet optimized for yield.

^{19}F NMR (188.31 MHz, MTBE/ toluene-d8): -121.40 (s, 24 F); -123.54 (s, 24 F); -222.84 (s, 12 F); ^{31}Al NMR: (104.27 MHz, 1,2-C$_2$H$_4$Cl$_2$/ toluene-d8): 64.7 (s)

IR: ṽ = 647.9 (m, δ_{C-F}), 954.2 (vs, ν_{C-C}), 968.7 (s, ν_{C-F}), 1249.5 (s, ν_{C-F}), 1267.6 (vs, ν_{C-F})

N.b. there are also absorptions at 1563.6 (vw), 1720.1 (w), and 2918.1 (vw) cm^{-1} which belong to an unknown organic impurity which could not be removed after washing with pentane or a 1:1 mixture of pentane/CH$_2$Cl$_2$ and each time drying for 3 d at 10^{-6} bar afterwards. Probably, this impurity (likely acetone) is strongly coordinated to the Na$^+$ cation.

MS (ESI$^-$, CH$_2$Cl$_2$): m/z = 421.1 (pfad$^-$); 534.8 (pfad$^-$ + 1,2-C$_6$H$_4$F$_2$); 842.7 (pfad$^-$ + pfad·); 1711.0 ([Al(pfad)$_4$]$^-$); 1739.8, 1740.9 ([Al(pfad)$_4$]$^-$ + N$_2$)

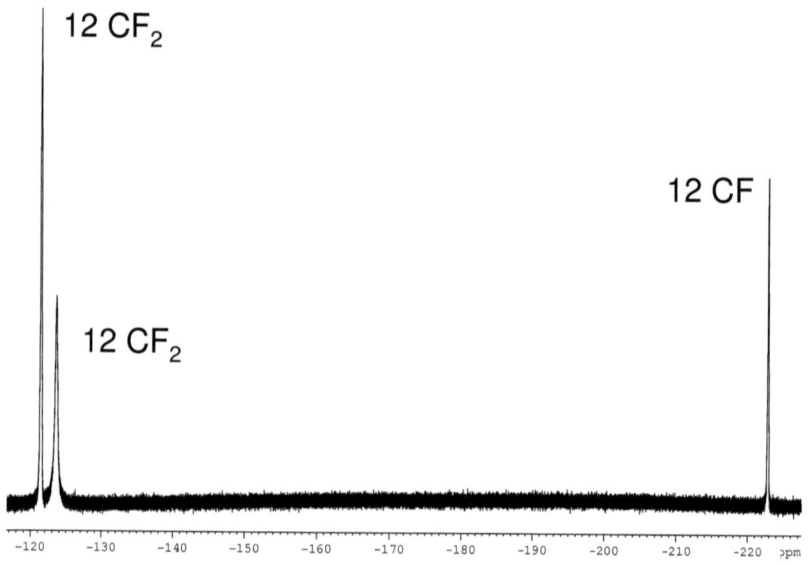

Figure 97: ^{19}F NMR of Na[Al(pfad)$_4$] in MTBE/ toluene-d8.

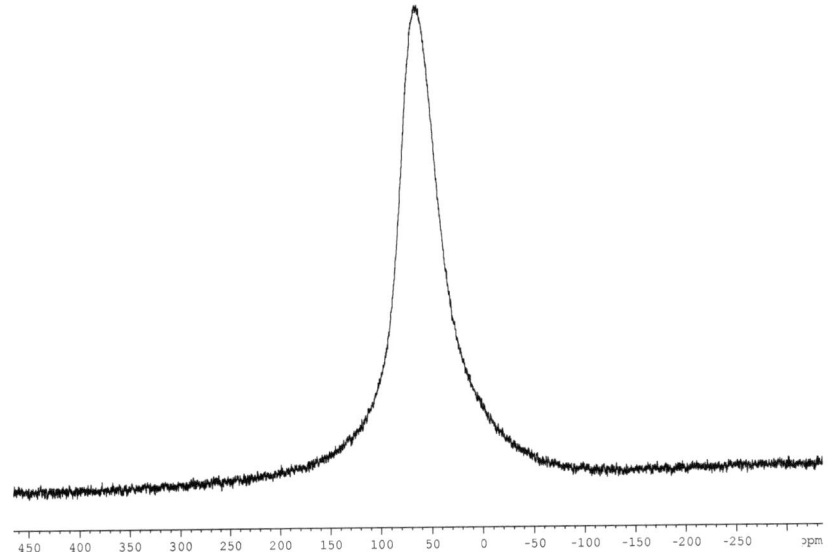

Figure 98: ^{27}Al NMR of Na[Al(pfad)$_4$] in 1,2-C$_2$H$_4$Cl$_2$/ toluene-d8.

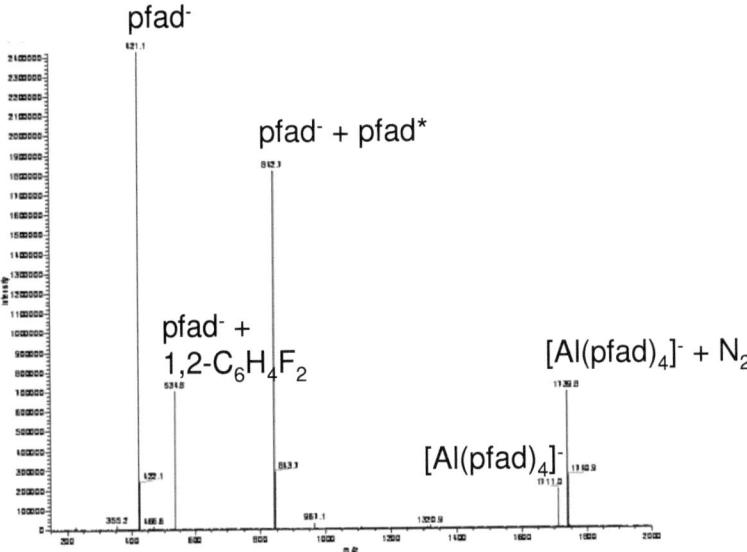

Figure 99: ESI$^-$ mass spectrum of Na[Al(pfad)$_4$].

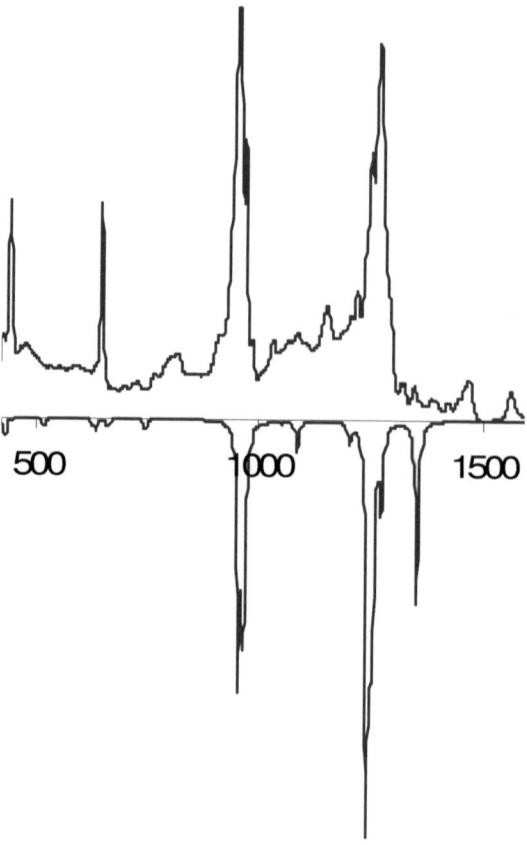

Figure 100: The IR absorption spectra of Na[Al(pfad)$_4$]. Top, mirrored: measured compound; bottom: calculated anion (BP86/SV(P)); both were scaled to the same maximum intensity.

5.2.6. Crystallization of ILs

[N$_{1,1,1,1}$][OTfa] (98%, iolitec), [N$_{4,4,4,4}$][BF$_4$] (99%, ABCR), [S$_{1,\varphi,\varphi}$][PF$_6$] (98%, iolitec), and [C$_2$MIm][OTos] (99%, iolitec) were crystallized from the commercially available materials. The latter compound (10 g) was dried in a 50 ml round bottom flask under vigorous stirring and constant vacuum (< 100 µbar) for 3 d at 70 °C, then allowed to cool to RT, forming an undercooled liquid phase which was then refilled through a funnel on air into another Schlenk flask to instantly form large, brown, hygroscopic crystals. For the remaining compounds, about 100 mg of each were separately dissolved in the minimum amount of DMF (Acros, 99%), then the triple volume of C$_2$H$_5$CN (Merck, 99.5%) was added and the solutions were allowed to slowly evaporate from glass vials in the fume hood at RT yielding small, clear crystals.

5.2.7. Crystal Structures

All crystal structures were measured and solved courtesy of B. Benkmil and Dr. G. Steinfeld, except for [S$_{1,\varphi,\varphi}$][PF$_6$] which was both measured and solved by A. Kraft. Additional help came from W. Beichel and Dr. G. Santiso-Quinoñes (all currently or formerly employed at the University of Freiburg).

Crystal data, structure refinement parameters, atomic coordinates, and equivalent isotropic displacement parameters for all crystallized compounds are given in the following.

Table 78: Crystal data and structure refinement parameters (I).

Identification code	[N$_{1,1,1,1}$][OTfa]	[N$_{4,4,4,4}$][BF$_4$]
Empirical formula	C$_6$H$_{12}$NO$_2$F$_3$	C$_{16}$H$_{36}$BNF$_4$
Formula weight	187.17	329.27
Temperature	100 K	100 K
Wavelength	0.71073 Å	0.71073 Å
Crystal system	Monoclinic	Monoclinic
Space group	P 2$_1$/m	P 2/c
Unit cell dimensions	a = 5.4059(11) Å, b = 8.4493(17) Å, c = 9.4910(19) Å; α = 90°, β = 93.69(3)°, γ = 90°	a = 33.6961(14) Å, b = 13.1536(5) Å, c = 26.8730(10) Å; α = 90°, β = 100.437(2)°, γ = 90°
Volume	432.61(15) Å^3	11713.7(8) Å^3
Z	2	24
Density (calculated)	1.437 Mg/m^3	1.120 Mg/m^3
Absorption coefficient	0.146 mm^{-1}	0.090 mm^{-1}
F(000)	196	4320
Crystal size	0.20x 0.19x 0.12 mm	0.08x 0.14x 0.24 mm
q range for data collection	3.23 to 27.46°	1.23 to 20.82°
Index ranges	-7 ≤ h ≤ 7, -10 ≤ k ≤ 10, -12 ≤ l ≤ 12	-33 ≤ h ≤ 33, -11 ≤ k ≤ 13, -26 ≤ l ≤ 25
Reflections collected	8193	34026
Independent reflections	1045 [R$_{int}$ = 0.0215]	12272 [R$_{int}$ = 0.0796]
Completeness	99.8% (θ = 27.48°)	100.0% (θ = 20.82°)
Absorption correction	None	Estimated minimum and maximum transmission: 0.7048 and 0.7447
Refinement method	Full-matrix least-squares on F^2	Full-matrix least-squares on F^2
Data / restraints / parameters	1045/ 0/ 92	12272/ 0/ 1190
Goodness-of-fit on F^2	S = 1.234	S = 1.018
R indices [for n reflections with I>2σ(I)]	R$_1$ = 0.0305, wR$_2$ = 0.0789 (n = 904)	R$_1$ = 0.0636, wR$_2$ = 0.1411 (n = 7654)
R indices (for all independent reflections)	R$_1$ = 0.0358, wR$_2$ = 0.0813	R$_1$ = 0.1135, wR$_2$ = 0.1701
Weighting scheme	w^{-1} = σ^2(F$_o^2$) + (aP)^2 + (bP), where P = [max(F$_o^2$, 0) + 2F$_c^2$]/3 a = 0.03, b = 0.3	w^{-1} = σ^2(F$_o^2$) + (aP)^2 + (bP), where P = [max(F$_o^2$, 0) + 2F$_c^2$]/3 a = 0.0675, b = 11.0385
Largest diff. peak and hole	0.325 and -0.181 eÅ^{-3}	0.704 and -0.592 eÅ^{-3}

Table 79: Crystal data and structure refinement parameters (II).

Identification code	[S$_{1,\varphi,\varphi}$][PF$_6$]	[C$_2$MIm][OTos]
Empirical formula	C$_{13}$H$_{13}$PSF$_6$	C$_{13}$H$_{18}$SN$_2$O$_3$
Formula weight	346.26	282.35
Temperature	150(2) K	100(2) K
Wavelength	0.71073 Å	0.71073 Å
Crystal system	Triclinic	Monoclinic
Space group	P $\bar{1}$	P 2$_1$/c
Unit cell dimensions	a = 9.1267(4) Å, b = 9.4796(4 Å, c = 10.4328(7) Å; α = 102.315(7)°, β = 102.237(7)°, γ = 116.843(8)°	a = 8.9584(18) Å, b = 30.874(6) Å, c = 10.112(2) Å; α = 90°, β = 90.94(3)°, γ = 90°
Volume	735.98(7) Å^3	2796.5(10) Å^3
Z	2	8
Density (calculated)	1.562 Mg/m^3	1.341 Mg/m^3
Absorption coefficient	0.383 mm^{-1}	0.237 mm^{-1}
F(000)	352	1200
Crystal size	0.25x 0.3x 0.1 mm	0.16x 0.11x 0.10 mm
q range for data collection	3.32 to 27.47°	3.01 to 27.48°
Index ranges	-11 ≤ h ≤ 11, -11 ≤ k ≤ 12, -13 ≤ l ≤ 13	-11 ≤ h ≤ 11, -40 ≤ k ≤ 40, -12 ≤ l ≤ 12
Reflections collected	28113	42851
Independent reflections	3362 [R$_{int}$ = 0.0327]	6290 [R$_{int}$ = 0.0564]
Completeness	99.7% (θ = 27.47°)	98.0% (θ = 27.48°)
Absorption correction	None	None
Refinement method	Full-matrix least-squares on F^2	Full-matrix least-squares on F^2
Data / restraints / parameters	3362/ 357/ 245	6290/ 0/ 475
Goodness-of-fit on F^2	S = 1.068	S = 1.076
R indices [for n reflections with I>2σ(I)]	R$_1$ = 0.0299, wR$_2$ = 0.0727 (n = 2930)	R$_1$ = 0.0389, wR$_2$ = 0.0885 (n = 4948)
R indices (for all independent reflections)	R$_1$ = 0.0357, wR$_2$ = 0.0755	R$_1$ = 0.0549, wR$_2$ = 0.0960
Weighting scheme	w^{-1} = σ^2(F$_o^2$) + (aP)^2 + (bP), where P = [max(F$_o^2$, 0) + 2F$_c^2$]/3 a = 0.0333, b = 0.3339	w^{-1} = σ^2(F$_o^2$) + (aP)^2 + (bP), where P = [max(F$_o^2$, 0) + 2F$_c^2$]/3 a = 0.0413, b = 1.2588
Largest diff. peak and hole	0.313 and -0.271 eÅ^{-3}	0.391 and -0.386 eÅ^{-3}

Table 80: Crystal data and structure refinement parameters (III).

Identification code	[Li$_2$(1,2-C$_6$H$_4$F$_2$)][Nb(hfip)$_6$]$_2$	[Ag(C$_6$H$_5$F)][Nb(hfip)$_6$]
Empirical formula	C$_{42}$H$_{16}$Li$_2$Nb$_2$ O$_{12}$F$_{74}$	C$_{24}$H$_{11}$NbAgO$_6$F$_{37}$
Formula weight	2318.25	1299.11
Temperature	115(2)K	110(2) K
Wavelength	0.71073 Å	0.71073 Å
Crystal system	Monoclinic	Monoclinic
Space group	C 2/c	P 2
Unit cell dimensions	a = 20.150(4)Å, b = 31.087(6)Å, c = 12.842(3)Å; α = 90°, β = 116.88(3)°, γ = 90°	a = 20.3402(6) Å, b = 18.9756(11) Å, c = 20.3669(11) Å; α = 90°, β = 105.541(6)°, γ = 90°
Volume	7175(2) Å^3	7573.6(6) Å^3
Z	4	8
Density (calculated)	2.146 Mg/m^3	2.279 Mg/m^3
Absorption coefficient	0.566 mm^{-1}	1.043 mm^{-1}
F(000)	4472	4992
Crystal size	0.23x 0.12x 0.06 mm	0.20 x 0.17 x 0.12 mm
q range for data collection	3.11 to 23.25°	3.01 to 26.05°
Index ranges	-22 ≤ h ≤ 22, -34 ≤ k ≤ 34, -14 ≤ l ≤ 14	-25≤h≤23, -23≤k≤23, -25≤l≤24
Reflections collected	74064	45942
Independent reflections	5145 [R$_{int}$ = 0.0377]	14563 [R$_{int}$ = 0.0557]
Completeness	99.7%	96.1 %
Absorption correction	Semi-empirical from equivalents 0.967 and 0.882	None
Refinement method	Full-matrix least-squares on F^2	Full-matrix least-squares on F^2
Data / restraints / parameters	5145/ 0/ 659	14563/ 0/ 1244
Goodness-of-fit on F^2	S = 1.045	S = 1.069
R indices [for n reflections with I>2σ(I)]	R$_1$ = 0.0303, wR$_2$ = 0.0661	R$_1$ = 0.0379, wR$_2$ = 0.0600
R indices (for all independent reflections)	R$_1$ = 0.0357, wR$_2$ = 0.0687	R$_1$ = 0.0485, wR$_2$ = 0.0638
Weighting scheme	w^{-1} = σ^2(F$_o^2$) + (aP)^2 + (bP), where P = [max(F$_o^2$, 0) + 2F$_c^2$]/3 a = 0.0223, b = 26.5	w^{-1} = σ^2(F$_o^2$) + (aP)^2 + (bP), where P = [max(F$_o^2$, 0) + 2F$_c^2$]/3 a = 0.002100, b = 15.363301
Largest diff. peak and hole	0.502 and -0.595 eÅ^{-3}	0.701 and -0.768 eÅ^{-3}

Table 81: Crystal data and structure refinement parameters (IV).

Identification code	[C$_4$MIm]$_2$[Fe$_2$(hfip)$_6$O]
Empirical formula	C$_{34}$H$_{36}$Fe$_2$N$_4$O$_7$F$_{36}$
Formula weight	1408.37
Temperature	110(2)K
Wavelength	0.71073 Å
Crystal system	Triclinic
Space group	P $\bar{1}$
Unit cell dimensions	a = 10.8516(4) Å, b = 12.8403(5) Å, c = 20.6854(7) Å; α = 90.245(2)°, β = 100.411(2)°, γ = 114.620(2)°
Volume	2566.87(16) Å^3
Z	2
Density (calculated)	1.822 Mg/m^3
Absorption coefficient	0.744 mm^{-1}
F(000)	1400
Crystal size	0.30x 0.30x 0.10 mm
q range for data collection	1.75 to 23.26°
Index ranges	-12 ≤ h ≤ 12, -14 ≤ k ≤ 14, -22 ≤ l ≤ 22
Reflections collected	33324
Independent reflections	7312 [R$_{int}$ = 0.0384]
Completeness	99.3%
Absorption correction	None 0.9293 and 0.8075
Refinement method	Full-matrix least-squares on F^2
Data / restraints / parameters	7312/ 645/ 1003
Goodness-of-fit on F^2	S = 1.087
R indices [for n reflections with I>2σ(I)]	R$_1$ = 0.0589, wR$_2$ = 0.1488
R indices (for all independent reflections)	R$_1$ = 0.0702, wR$_2$ = 0.1555
Weighting scheme	w^{-1} = σ^2(F$_o^2$) + (aP)^2 + (bP), where P = [max(F$_o^2$, 0) + 2F$_c^2$]/3 a = 0.059, b = 9
Largest diff. peak and hole	0.909 and -0.558 eÅ^{-3}

Table 82: Atomic coordinates (x 10^4) and equivalent isotropic displacement parameters (Å^2 x 10^3) for [C$_4$MIm]$_2$[Fe$_2$(hfip)$_6$O]. U(eq) is defined as one third of the trace of the orthogonalized U$_{ij}$ tensor.

	x	y	z	U(eq)
Fe(1)	6447(1)	982(1)	2784(1)	24(1)
Fe(2)	6708(1)	3822(1)	2737(1)	25(1)
O(3)	6862(4)	546(3)	2006(2)	36(1)
O(4)	5691(3)	4021(3)	1929(2)	36(1)
O(5)	6027(4)	4228(3)	3409(2)	47(1)
O(6)	8562(3)	4883(3)	2816(2)	39(1)
O(7)	6609(3)	2422(3)	2777(2)	29(1)
O(1)	4627(3)	-22(3)	2840(2)	39(1)
C(1)	3541(5)	262(4)	2853(3)	39(1)
C(2)	2449(7)	-263(5)	2247(4)	66(2)
C(3)	3010(7)	-101(6)	3491(4)	63(2)
F(1)	1925(4)	-1408(3)	2188(3)	111(2)
F(2)	1393(4)	15(4)	2204(3)	97(2)
F(3)	3002(5)	93(4)	1710(2)	89(1)
F(4)	1872(4)	70(4)	3499(3)	98(2)
F(5)	2640(5)	-1228(4)	3538(3)	108(2)
F(6)	3951(5)	471(5)	4004(2)	89(1)

C(7)	7343(6)	1271(4)	1538(3)	39(1)
C(8)	8869(6)	1633(5)	1578(3)	48(1)
C(9)	6505(7)	673(6)	866(3)	56(2)
C(10)	4789(6)	3116(5)	1456(3)	48(1)
C(11)	3289(7)	2847(6)	1460(4)	66(2)
C(12)	5133(7)	3431(7)	786(3)	66(2)
C(16)	9679(5)	4633(4)	2819(3)	36(1)
C(17)	10248(7)	5020(5)	2199(3)	53(2)
C(18)	10763(6)	5197(5)	3442(4)	58(2)
O(2)	7629(4)	771(3)	3498(2)	41(1)
C(4A)	8570(30)	1739(15)	4027(14)	29(4)
C(5A)	9940(30)	1845(18)	4100(20)	35(5)
C(6A)	8142(19)	1689(19)	4641(8)	52(4)
F(7A)	10110(20)	881(18)	4122(15)	70(6)
F(8A)	10311(17)	2291(12)	3593(10)	74(4)
F(9A)	10870(30)	2490(30)	4584(15)	92(8)
F(10A)	8090(30)	670(20)	4961(14)	101(7)
F(11A)	8918(16)	2538(14)	5079(5)	88(5)
F(12A)	6815(16)	1560(19)	4548(6)	113(6)
C(4B)	8350(30)	1350(16)	4011(15)	35(4)
C(5B)	9860(40)	1450(20)	4112(19)	40(7)
C(6B)	7720(20)	930(40)	4635(11)	80(7)
F(7B)	9850(30)	490(20)	4256(15)	82(7)
F(8B)	10450(18)	1770(20)	3584(8)	103(7)
F(9B)	10600(30)	2250(30)	4656(13)	94(9)
F(10B)	7784(15)	-196(19)	4660(8)	135(8)
F(11B)	8255(18)	1430(30)	5144(7)	126(11)
F(12B)	6376(14)	540(30)	4476(7)	175(12)
F(13)	9188(4)	779(3)	1445(2)	61(1)
F(14)	9392(4)	2429(3)	1163(2)	69(1)
F(15)	9564(4)	2105(3)	2184(2)	63(1)
F(16)	6509(4)	-342(3)	746(2)	71(1)
F(17)	6927(5)	1306(4)	369(2)	81(1)
F(18)	5178(4)	478(4)	836(2)	81(1)
F(19)	2990(5)	3748(4)	1340(3)	109(2)
F(20)	2398(4)	2006(4)	1027(2)	90(1)
F(21)	3034(4)	2568(4)	2060(2)	90(1)
F(22)	4343(5)	2627(4)	297(2)	95(2)
F(23)	5057(5)	4400(4)	615(2)	99(2)
F(24)	6448(5)	3603(5)	794(2)	103(2)
C(13)	5760(30)	3500(20)	4036(13)	41(5)
C(14)	4300(30)	3250(30)	4128(15)	44(5)
F(25)	4070(20)	4148(18)	4190(13)	77(6)
F(26)	3950(20)	2580(18)	4607(11)	80(5)
F(27)	3340(30)	2600(30)	3534(19)	66(6)
C(15)	6910(20)	4180(30)	4602(12)	61(6)
F(28)	6950(20)	5160(30)	4745(17)	122(12)
F(29)	6668(17)	3570(30)	5127(7)	112(8)
F(30)	8030(20)	4190(30)	4486(11)	92(6)
C(13B)	5750(30)	3810(20)	3894(12)	35(4)
C(14B)	4290(30)	3540(30)	3953(14)	44(6)
F(25B)	3480(30)	2900(30)	3478(18)	84(9)
F(26B)	4080(20)	4500(20)	3985(10)	73(5)
F(27B)	3936(15)	3040(20)	4519(9)	57(4)
C(15B)	6820(20)	4690(30)	4513(11)	47(5)
F(28B)	6760(20)	5638(16)	4554(10)	81(6)
F(29B)	6580(16)	4194(15)	5100(6)	67(4)
F(30B)	8103(19)	4810(20)	4481(9)	85(6)
F(31)	11266(4)	4718(3)	2142(2)	76(1)
F(32)	10734(4)	6151(3)	2176(2)	79(1)
F(33)	9256(5)	4544(4)	1674(2)	81(1)

F(34)	11109(4)	6322(3)	3525(3)	94(2)
F(35)	11921(4)	5077(4)	3438(2)	86(1)
F(36)	10299(5)	4742(4)	3966(2)	83(1)
N(1)	7978(5)	7477(3)	1974(2)	42(1)
N(2)	8576(4)	7895(4)	3013(2)	43(1)
C(50)	8434(5)	7152(4)	2546(3)	45(1)
C(51)	7824(5)	8460(4)	2106(3)	43(1)
C(52)	8208(5)	8728(4)	2755(3)	42(1)
C(53)	7622(7)	6824(5)	1320(4)	71(2)
C(54)	8773(8)	7199(6)	952(4)	68(2)
C(55)	9244(7)	8393(6)	732(3)	62(2)
C(56)	10405(8)	8690(7)	347(4)	83(2)
C(57)	9056(7)	7839(6)	3721(3)	68(2)
N(3)	4365(15)	7072(10)	1942(6)	59(3)
N(4)	4727(16)	7007(13)	2985(6)	51(3)
C(60)	4431(10)	7583(8)	2519(6)	49(2)
C(61)	4674(12)	6145(9)	2071(6)	52(3)
C(62)	4921(10)	6089(8)	2713(6)	41(2)
C(63)	4146(17)	7538(14)	1300(9)	90(4)
C(64)	2738(18)	6978(14)	897(8)	83(4)
C(65)	2234(19)	5721(14)	666(10)	88(5)
C(66)	809(17)	5253(17)	214(8)	117(6)
C(67)	5190(40)	7500(30)	3667(11)	67(6)
N(5)	4810(20)	6838(15)	2760(6)	46(4)
N(6)	4980(40)	7490(30)	3752(9)	63(5)
C(70)	4754(15)	7650(11)	3127(7)	48(3)
C(71)	5080(14)	6113(12)	3176(7)	53(3)
C(72)	5393(16)	6588(13)	3782(7)	68(4)
C(73)	4260(20)	6570(15)	2047(7)	54(4)
C(74)	3964(19)	7444(16)	1684(8)	67(4)
C(75)	3700(20)	7265(18)	934(8)	86(5)
C(76)	2340(30)	6320(20)	645(14)	122(12)
C(77)	5406(19)	8354(14)	4276(8)	72(4)

Table 83: Atomic coordinates (x 10^4) and equivalent isotropic displacement parameters (Å^2 x 10^3) for [Li$_2$(1,2-C$_6$H$_4$F$_2$)][Nb(hfip)$_6$]$_2$. U(eq) is defined as one third of the trace of the orthogonalized U$_{ij}$ tensor.

	x	y	z	U(eq)
Nb(1)	2348(1)	8802(1)	11850(1)	21(1)
O(1)	2528(1)	8766(1)	10520(2)	28(1)
O(2)	1421(1)	8412(1)	11068(2)	27(1)
O(3)	1731(1)	9296(1)	11299(2)	33(1)
O(4)	2945(1)	8284(1)	12439(2)	27(1)
O(5)	2056(1)	8671(1)	13154(2)	25(1)
O(6)	3146(1)	9158(1)	12859(2)	36(1)
C(1)	2880(2)	8548(1)	9954(3)	31(1)
C(2)	988(2)	8268(1)	9930(3)	29(1)
C(3)	1163(2)	9582(1)	10656(3)	35(1)
C(4)	3339(2)	8092(1)	13523(3)	29(1)
C(5)	1914(2)	8936(1)	13902(2)	28(1)
C(11)	2446(2)	8619(1)	8652(3)	40(1)
C(12)	3683(2)	8683(2)	10402(3)	58(1)
C(21)	172(2)	8343(1)	9624(3)	41(1)
C(22)	1158(2)	7794(1)	9870(3)	38(1)
C(31)	1140(2)	9640(1)	9460(3)	47(1)
C(32)	1301(2)	10005(1)	11320(3)	50(1)
C(41)	4155(2)	8059(1)	13790(3)	41(1)
C(42)	3001(2)	7662(1)	13544(4)	52(1)
C(51)	2326(2)	8758(1)	15142(3)	50(1)
C(52)	1075(2)	8947(1)	13480(3)	38(1)

	x	y	z	U(eq)
F(11)	2786(1)	8436(1)	8079(2)	54(1)
F(12)	2346(1)	9029(1)	8362(2)	62(1)
F(13)	1775(1)	8440(1)	8250(2)	67(1)
F(14)	4027(1)	8695(1)	11558(2)	78(1)
F(15)	3776(2)	9050(1)	10032(2)	104(1)
F(16)	4033(1)	8388(1)	10043(2)	99(1)
F(21)	42(1)	8762(1)	9662(2)	53(1)
F(22)	-20(1)	8151(1)	10376(2)	60(1)
F(23)	-287(1)	8204(1)	8564(2)	72(1)
F(24)	732(1)	7616(1)	8846(2)	59(1)
F(25)	1868(1)	7744(1)	10076(2)	46(1)
F(26)	1081(1)	7564(1)	10695(2)	49(1)
F(31)	634(1)	9930(1)	8819(2)	61(1)
F(32)	1795(1)	9759(1)	9540(2)	72(1)
F(33)	966(2)	9273(1)	8879(2)	68(1)
F(34)	716(2)	10268(1)	10825(2)	71(1)
F(35)	1434(2)	9935(1)	12404(2)	75(1)
F(36)	1887(2)	10217(1)	11365(3)	77(1)
F(41)	4549(1)	7850(1)	14798(2)	56(1)
F(42)	4439(1)	8454(1)	13905(2)	56(1)
F(43)	4256(1)	7857(1)	12971(2)	61(1)
F(44)	2258(1)	7707(1)	13217(2)	56(1)
F(45)	3047(2)	7379(1)	12825(3)	91(1)
F(46)	3294(2)	7493(1)	14619(3)	92(1)
F(51)	3042(1)	8713(1)	15444(2)	68(1)
F(52)	2076(2)	8368(1)	15225(2)	83(1)
F(53)	2271(2)	9013(1)	15922(2)	82(1)
F(54)	794(1)	8543(1)	13235(2)	58(1)
F(55)	737(1)	9168(1)	12493(2)	48(1)
F(56)	873(1)	9113(1)	14242(2)	53(1)
C(61)	3862(3)	9580(2)	14551(4)	73(2)
F(62)	3348(2)	9579(1)	14887(2)	93(1)
C(6)	3412(5)	9574(3)	13164(8)	46(2)
C(62)	3894(9)	9691(3)	12622(9)	68(4)
F(61)	4125(4)	10003(2)	14831(6)	85(2)
F(63)	4355(4)	9328(2)	14979(5)	79(2)
F(64)	4056(8)	10119(4)	12796(12)	129(6)
F(65)	3514(5)	9627(3)	11452(9)	82(3)
F(66)	4520(3)	9479(2)	12985(4)	90(2)
C(6B)	3746(9)	9425(6)	13347(15)	58(5)
C(62B)	3651(12)	9860(7)	12590(20)	68(7)
F(61B)	4463(8)	9775(5)	15085(11)	114(5)
F(63B)	4186(7)	9109(4)	15176(9)	90(4)
F(64B)	4262(10)	10019(8)	12780(17)	83(6)
F(65B)	3269(9)	9774(6)	11504(18)	75(5)
F(66B)	3232(5)	10121(3)	12903(8)	96(3)
Li(1)	1379(3)	8164(2)	12423(5)	40(1)
F(1)	699(1)	7722(1)	12642(2)	42(1)
C(71)	356(2)	7335(1)	12566(3)	29(1)
C(72)	722(2)	6959(1)	12618(3)	39(1)
C(73)	356(2)	6576(1)	12556(3)	49(1)

Table 84: Atomic coordinates (x 10^4) and equivalent isotropic displacement parameters (Å^2 x 10^3) for [N$_{4,4,4,4}$][BF$_4$]. U(eq) is defined as one third of the trace of the orthogonalized U$_{ij}$ tensor.

	x	y	z	U(eq)		x	y	z	U(eq)
B(1)	3327(2)	497(5)	948(3)	37(2)	C(243)	4355(2)	10350(4)	2519(2)	51(2)
B(2)	2240(2)	3415(5)	3253(2)	32(2)	C(244)	4157(2)	11392(5)	2501(2)	63(2)
B(3)	4463(2)	7047(5)	3585(2)	42(2)	C(311)	2469(1)	9123(5)	-181(2)	27(1)

B(4)	0	7026(7)	2500	36(2)	C(312)	2432(2)	9356(3)	-741(2)	29(1)
B(5)	3266(2)	5378(5)	833(2)	38(2)	C(313)	2323(2)	10470(3)	-834(2)	33(1)
B(6)	1206(2)	9190(5)	159(2)	33(2)	C(314)	2321(2)	10785(4)	-1377(2)	55(2)
B(7)	0	1368(6)	2500	24(2)	C(321)	2120(1)	7449(3)	-215(2)	24(1)
F(11)	3480(1)	1305(2)	1254(1)	41(1)	C(322)	2098(1)	6389(3)	2(2)	29(1)
F(12)	3214(1)	-263(3)	1244(1)	77(1)	C(323)	1706(2)	5876(4)	-222(2)	33(1)
F(13)	3022(1)	831(3)	591(2)	109(2)	C(324)	1665(2)	4829(4)	5(2)	37(1)
F(14)	3620(1)	97(2)	713(1)	67(1)	C(331)	2841(1)	7501(4)	-270(2)	27(1)
F(21)	1937(1)	3434(2)	2823(1)	36(1)	C(332)	3256(1)	7996(4)	-173(2)	31(1)
F(22)	2089(1)	3778(2)	3664(1)	49(1)	C(333)	3536(2)	7414(4)	-463(2)	36(1)
F(23)	2562(1)	4002(2)	3161(1)	46(1)	C(334)	3412(2)	7424(4)	-1035(2)	46(2)
F(24)	2368(1)	2415(2)	3344(1)	62(1)	C(341)	2633(1)	7960(4)	538(2)	28(1)
F(31)	4110(1)	6517(2)	3576(1)	47(1)	C(342)	2321(2)	8377(4)	828(2)	32(1)
F(32)	4387(1)	8070(3)	3554(1)	75(1)	C(343)	2448(2)	8199(4)	1390(2)	34(1)
F(33)	4731(1)	6813(3)	4023(1)	65(1)	C(344)	2145(2)	8624(4)	1689(2)	42(2)
F(34)	4630(1)	6765(3)	3169(1)	72(1)	C(411)	3100(1)	3390(4)	1868(2)	31(1)
F(41)	75(1)	6423(3)	2115(1)	82(1)	C(412)	3289(2)	3655(4)	2408(2)	35(1)
F(42)	332(1)	7642(3)	2647(1)	72(1)	C(413)	3711(2)	4093(4)	2422(2)	39(1)
F(51)	3503(1)	4536(2)	980(1)	41(1)	C(414)	3944(2)	4224(4)	2958(2)	50(2)
F(52)	3154(1)	5835(2)	1246(1)	52(1)	C(421)	2569(2)	2647(4)	1238(2)	36(1)
F(53)	3486(1)	6081(2)	608(1)	77(1)	C(422)	2132(2)	2358(5)	1045(2)	49(2)
F(54)	2925(1)	5081(3)	510(1)	91(1)	C(423)	2071(2)	2056(5)	481(2)	62(2)
F(61)	1490(1)	9484(2)	-130(1)	35(1)	C(424)	1632(2)	1906(5)	254(2)	74(2)
F(62)	1204(1)	8136(2)	187(1)	60(1)	C(431)	2610(2)	2199(3)	2159(2)	31(1)
F(63)	1305(1)	9583(2)	642(1)	48(1)	C(432)	2871(2)	1260(4)	2162(2)	36(1)
F(64)	829(1)	9528(2)	-61(1)	50(1)	C(433)	2814(2)	570(4)	2598(2)	40(2)
F(71)	208(1)	1987(2)	2211(1)	36(1)	C(434)	3081(2)	-357(4)	2639(2)	48(2)
F(72)	270(1)	761(2)	2819(1)	32(1)	C(441)	2380(2)	3881(4)	1873(2)	30(1)
N(1)	3839(1)	7146(3)	4888(2)	41(1)	C(442)	2357(2)	4782(4)	1516(2)	33(1)
N(2)	4150(1)	7754(3)	1828(1)	26(1)	C(443)	2001(2)	5469(4)	1576(2)	33(1)
N(3)	2518(1)	8009(3)	-30(1)	24(1)	C(444)	2037(2)	5918(4)	2105(2)	40(2)
N(4)	2663(1)	3026(3)	1782(2)	28(1)	C(511)	301(1)	10031(3)	831(2)	23(1)
N(5)	345(1)	9236(3)	1250(1)	20(1)	C(512)	82(2)	11006(3)	932(2)	26(1)
N(6)	923(1)	4403(3)	3087(1)	25(1)	C(513)	56(2)	11720(3)	488(2)	28(1)
C(111)	3978(2)	7572(4)	5419(2)	39(2)	C(514)	-171(2)	12688(4)	559(2)	33(1)
C(112)	4423(2)	7838(4)	5555(2)	45(2)	C(521)	587(1)	8369(3)	1072(2)	21(1)
C(113)	4505(2)	8403(5)	6053(2)	68(2)	C(522)	649(1)	7454(3)	1421(2)	25(1)
C(114)	4955(2)	8568(5)	6248(3)	93(3)	C(523)	967(1)	6733(4)	1274(2)	28(1)
C(121)	4103(2)	6262(4)	4777(2)	43(2)	C(524)	834(2)	6209(4)	767(2)	33(1)
C(122)	4090(2)	5333(4)	5100(2)	52(2)	C(531)	-64(1)	8882(3)	1342(2)	23(1)
C(123)	4435(2)	4614(5)	5056(2)	65(2)	C(532)	-346(1)	8453(4)	888(2)	27(1)
C(124)	4446(2)	4261(4)	4514(2)	66(2)	C(533)	-739(1)	8069(3)	1034(2)	29(1)
C(131)	3870(2)	7945(4)	4486(2)	43(2)	C(534)	-1000(2)	8928(4)	1161(2)	53(2)
C(132)	3600(2)	8871(5)	4487(2)	66(2)	C(541)	564(1)	9668(3)	1747(2)	21(1)
C(133)	3632(2)	9556(5)	4027(2)	55(2)	C(542)	993(1)	10025(4)	1750(2)	28(1)
C(134)	4029(2)	10101(5)	4100(3)	70(2)	C(543)	1149(1)	10628(4)	2228(2)	28(1)
C(141)	3405(2)	6818(5)	4867(2)	65(2)	C(544)	1584(1)	10957(4)	2262(2)	38(1)
C(142)	3192(2)	6414(7)	4360(2)	118(4)	C(611)	1260(1)	5177(3)	3122(2)	23(1)
C(143)	2763(3)	6122(14)	4389(4)	308(14)	C(612)	1147(1)	6191(3)	2857(2)	28(1)
C(144)	2492(3)	6214(7)	4106(5)	183(7)	C(613)	1478(2)	6967(3)	3030(2)	30(1)
C(211)	3726(1)	7519(4)	1907(2)	25(1)	C(614)	1401(2)	7959(3)	2727(2)	36(1)
C(212)	3674(1)	7242(4)	2438(2)	30(1)	C(621)	1117(1)	3445(3)	3341(2)	26(1)
C(213)	3242(1)	6889(4)	2439(2)	33(1)	C(622)	839(2)	2550(4)	3364(2)	35(1)
C(214)	3158(2)	6682(4)	2967(2)	45(2)	C(623)	1031(2)	1789(4)	3752(2)	47(2)
C(221)	4116(1)	8064(4)	1280(2)	28(1)	C(624)	1396(2)	1339(4)	3633(2)	43(2)
C(222)	4504(2)	8377(4)	1109(2)	34(1)	C(631)	590(1)	4786(4)	3349(2)	30(1)
C(223)	4427(2)	8515(4)	540(2)	34(1)	C(632)	701(1)	5000(4)	3906(2)	27(1)
C(224)	4795(2)	8861(5)	338(2)	61(2)	C(633)	342(2)	5410(4)	4109(2)	31(1)
C(231)	4427(2)	6838(4)	1949(2)	36(1)	C(634)	436(2)	5645(4)	4673(2)	38(1)
C(232)	4331(2)	5914(4)	1618(2)	43(2)	C(641)	726(1)	4199(4)	2539(2)	27(1)
C(233)	4606(2)	5047(4)	1859(2)	62(2)	C(642)	1000(1)	3838(4)	2191(2)	27(1)

C(234)	4597(2)	4166(5)	1539(3)	100(3)	C(643)	758(2)	3699(4)	1660(2)	35(1)
C(241)	4342(1)	8586(4)	2181(2)	32(1)	C(644)	1003(2)	3314(4)	1279(2)	45(2)
C(242)	4130(2)	9606(4)	2133(2)	35(1)					

Table 85: Atomic coordinates (x 10^4) and equivalent isotropic displacement parameters (Å^2 x 10^3) for [Ag(C$_6$H$_5$F)][Nb(hfip)$_6$]. U(eq) is defined as one third of the trace of the orthogonalized U$_{ij}$ tensor.

	x	y	z	U(eq)		x	y	z	U(eq)
Ag(1)	6459(1)	4059(1)	4070(1)	30(1)	Ag(2)	1593(1)	1074(1)	4261(1)	22(1)
Nb(1)	5717(1)	2656(1)	3231(1)	12(1)	Nb(2)	657(1)	2329(1)	3331(1)	12(1)
F(101)	8617(2)	4541(2)	5557(2)	41(1)	F(201)	3820(2)	667(2)	5685(2)	42(1)
F(111)	5138(2)	3141(2)	4769(1)	34(1)	F(211)	518(2)	603(2)	2008(2)	35(1)
F(112)	4276(2)	3818(2)	4667(2)	43(1)	F(212)	1479(2)	372(2)	1809(2)	44(1)
F(113)	5267(2)	4256(2)	4744(2)	47(1)	F(213)	1309(2)	101(2)	2777(2)	38(1)
F(114)	4238(2)	4261(2)	2671(2)	47(1)	F(214)	2502(2)	1269(2)	2297(2)	35(1)
F(115)	3872(2)	4577(2)	3523(2)	48(1)	F(215)	2405(1)	1993(2)	3070(2)	31(1)
F(116)	4878(2)	4882(2)	3469(2)	52(1)	F(216)	2474(1)	888(2)	3284(2)	37(1)
F(121)	7496(1)	3112(2)	2589(2)	30(1)	F(221)	1016(1)	2585(2)	5226(1)	24(1)
F(122)	7348(2)	4024(2)	1943(2)	45(1)	F(222)	1970(2)	2032(1)	5553(1)	25(1)
F(123)	7467(2)	4136(2)	3018(2)	40(1)	F(223)	1922(2)	3143(2)	5764(1)	27(1)
F(124)	5375(2)	4248(2)	2096(2)	34(1)	F(224)	2696(1)	3081(2)	4095(1)	28(1)
F(125)	6123(2)	4625(2)	1613(2)	43(1)	F(225)	2951(1)	3190(2)	5184(1)	29(1)
F(126)	6231(2)	4833(2)	2670(2)	44(1)	F(226)	2840(1)	2161(1)	4726(1)	25(1)
F(131)	8061(1)	2144(2)	4925(2)	35(1)	F(231)	326(2)	933(2)	5077(2)	40(1)
F(132)	7638(1)	2564(2)	3922(1)	26(1)	F(232)	-764(2)	860(2)	4787(2)	43(1)
F(133)	7760(2)	3233(2)	4800(2)	33(1)	F(233)	-280(2)	1874(2)	4945(2)	34(1)
F(134)	6768(2)	3119(2)	5491(2)	33(1)	F(234)	259(2)	149(2)	3925(2)	39(1)
F(135)	7271(2)	2124(2)	5773(1)	36(1)	F(235)	-253(2)	625(2)	2972(2)	31(1)
F(136)	6189(2)	2162(2)	5326(1)	30(1)	F(236)	-830(2)	206(2)	3622(2)	36(1)
F(141)	4836(2)	1966(2)	1126(2)	38(1)	F(241)	-931(2)	3742(2)	3116(2)	34(1)
F(142)	5219(2)	3010(2)	1065(2)	34(1)	F(242)	-1270(1)	3656(2)	4028(2)	36(1)
F(143)	4163(2)	2786(2)	614(1)	36(1)	F(243)	-1080(1)	2720(2)	3506(1)	27(1)
F(144)	3807(2)	2674(2)	2515(1)	35(1)	F(244)	-206(2)	4248(2)	4881(2)	44(1)
F(145)	3340(2)	2765(2)	1440(2)	45(1)	F(245)	791(1)	4035(1)	4751(1)	25(1)
F(146)	3887(2)	1822(2)	1858(2)	46(1)	F(246)	52(2)	4567(1)	3965(2)	37(1)
F(151)	7536(2)	638(2)	2988(2)	44(1)	F(251)	-452(2)	2969(2)	1378(2)	41(1)
F(152)	7240(1)	1031(1)	3850(1)	26(1)	F(252)	-717(2)	2031(2)	786(2)	41(1)
F(153)	6535(2)	423(2)	3095(2)	36(1)	F(253)	304(1)	2176(2)	1408(1)	33(1)
F(154)	6339(2)	2106(1)	1728(1)	26(1)	F(254)	-1462(2)	2657(2)	1984(2)	37(1)
F(155)	6963(2)	1181(2)	1784(2)	47(1)	F(255)	-1403(2)	1661(2)	2493(2)	35(1)
F(156)	5939(2)	1092(2)	1878(2)	42(1)	F(256)	-1678(2)	1701(2)	1401(2)	42(1)
F(161)	3908(2)	1335(2)	3264(2)	35(1)	F(261)	2028(2)	3860(2)	1883(2)	62(1)
F(162)	3968(2)	1202(2)	4324(2)	37(1)	F(262)	1701(2)	2785(2)	1824(2)	40(1)
F(163)	3925(1)	2244(2)	3897(2)	31(1)	F(263)	971(2)	3621(2)	1544(2)	54(1)
F(164)	5954(1)	961(2)	4235(2)	30(1)	F(264)	1657(2)	4695(2)	2777(2)	45(1)
F(165)	5095(2)	451(2)	4451(2)	52(1)	F(265)	1395(2)	4166(2)	3603(1)	37(1)
F(166)	5111(2)	624(2)	3415(2)	42(1)	F(266)	652(2)	4301(2)	2648(2)	47(1)
O(11)	5313(2)	3518(2)	3533(2)	19(1)	O(21)	1147(2)	1483(2)	3086(2)	16(1)
O(12)	6307(2)	3368(2)	2955(2)	19(1)	O(22)	1504(2)	2328(2)	4127(1)	16(1)
O(13)	6397(2)	2803(2)	4152(2)	17(1)	O(23)	331(2)	1641(2)	3904(2)	19(1)
O(14)	5089(2)	2665(2)	2354(2)	20(1)	O(24)	287(2)	3098(2)	3721(2)	16(1)
O(15)	6253(2)	1894(2)	3049(2)	16(1)	O(25)	-116(2)	2172(2)	2585(2)	19(1)
O(16)	5163(2)	2066(2)	3609(2)	17(1)	O(26)	1078(2)	2941(2)	2830(2)	18(1)
C(11)	4715(3)	3676(3)	3703(3)	22(1)	C(21)	1440(2)	1324(2)	2558(2)	17(1)
C(12)	6426(2)	3657(2)	2363(2)	18(1)	C(22)	1828(2)	2817(2)	4620(2)	17(1)
C(13)	6874(2)	2377(2)	4585(2)	16(1)	C(23)	-246(2)	1275(3)	3948(2)	20(1)
C(14)	4539(2)	2856(3)	1815(2)	22(1)	C(24)	-119(2)	3347(2)	4123(2)	17(1)
C(15)	6767(2)	1595(2)	2803(2)	19(1)	C(25)	-528(2)	1903(2)	1971(2)	20(1)
C(16)	4957(2)	1623(2)	4061(2)	17(1)	C(26)	1516(3)	3458(3)	2715(3)	27(1)
C(101)	7998(3)	4754(3)	5197(3)	30(1)	C(201)	2663(3)	605(3)	5626(3)	27(1)
C(102)	7468(3)	4712(3)	5497(3)	33(1)	C(202)	2014(3)	355(3)	5303(3)	27(1)

C(103)	6825(3)	4932(3)	5136(3)	36(2)	C(203)	1914(3)	-53(3)	4720(3)	28(1)
C(104)	6724(3)	5194(3)	4474(3)	41(2)	C(204)	2468(3)	-205(3)	4460(3)	31(1)
C(105)	7272(3)	5222(3)	4188(3)	39(2)	C(205)	3109(3)	51(3)	4783(3)	31(1)
C(106)	7915(3)	5002(3)	4555(3)	34(1)	C(206)	3182(3)	441(3)	5363(3)	26(1)
C(111)	4846(3)	3732(3)	4476(3)	27(1)	C(211)	1190(3)	589(3)	2289(3)	26(1)
C(112)	4426(3)	4362(3)	3349(3)	32(1)	C(212)	2214(3)	1362(3)	2806(2)	21(1)
C(121)	7189(3)	3737(3)	2476(3)	24(1)	C(221)	1689(2)	2644(3)	5299(2)	18(1)
C(122)	6045(3)	4345(3)	2187(3)	24(1)	C(222)	2584(2)	2804(3)	4662(2)	19(1)
C(131)	7593(2)	2580(3)	4566(2)	20(1)	C(231)	-239(3)	1230(3)	4696(3)	26(1)
C(132)	6781(3)	2447(3)	5308(2)	25(1)	C(232)	-268(3)	552(3)	3617(3)	27(1)
C(141)	4685(3)	2643(3)	1147(2)	25(1)	C(241)	-856(3)	3376(3)	3687(3)	24(1)
C(142)	3892(3)	2516(3)	1905(3)	30(1)	C(242)	131(2)	4058(3)	4423(2)	21(1)
C(151)	7015(3)	912(3)	3179(2)	21(1)	C(251)	-359(3)	2278(3)	1381(3)	24(1)
C(152)	6497(3)	1483(3)	2042(3)	26(1)	C(252)	-1270(3)	1985(3)	1963(3)	29(1)
C(161)	4178(3)	1598(3)	3880(3)	24(1)	C(261)	1561(3)	3434(3)	1994(3)	35(1)
C(162)	5273(3)	903(3)	4035(3)	27(1)	C(262)	1292(3)	4175(3)	2925(3)	27(1)

Table 86: Atomic coordinates (x 10^4) and equivalent isotropic displacement parameters ($Å^2$ x 10^3) for [C$_2$MIm][OTos]. U(eq) is defined as one third of the trace of the orthogonalized U$_{ij}$ tensor.

	x	y	z	U(eq)		x	y	z	U(eq)
S(1)	10309(1)	4104(1)	6283(1)	17(1)	C(8)	9580(2)	3243(1)	6518(2)	22(1)
S(2)	4483(1)	5969(1)	8070(1)	17(1)	C(9)	8897(2)	2859(1)	6119(2)	26(1)
O(1)	9854(1)	4435(1)	5339(1)	22(1)	C(10)	8108(2)	2834(1)	4918(2)	23(1)
O(2)	11909(2)	4034(1)	6328(2)	35(1)	C(11)	8057(2)	3199(1)	4109(2)	22(1)
O(3)	9675(2)	4172(1)	7584(1)	28(1)	C(12)	8743(2)	3585(1)	4497(2)	18(1)
O(4)	4050(1)	5712(1)	6916(1)	25(1)	C(13)	7302(3)	2424(1)	4513(3)	38(1)
O(5)	5576(1)	5745(1)	8910(1)	20(1)	C(14)	3561(2)	5689(1)	3912(2)	17(1)
O(6)	3237(1)	6136(1)	8814(1)	28(1)	C(15)	4045(2)	5570(1)	1810(2)	20(1)
N(1)	10326(2)	4219(1)	1005(1)	17(1)	C(16)	3799(2)	5183(1)	2396(2)	19(1)
N(2)	8880(2)	4771(1)	1298(1)	16(1)	C(17)	3957(2)	6355(1)	2562(2)	26(1)
N(3)	3503(2)	5264(1)	3710(1)	16(1)	C(18)	5405(2)	6493(1)	1955(2)	29(1)
N(4)	3893(2)	5883(1)	2776(2)	19(1)	C(19)	3108(2)	4939(1)	4706(2)	20(1)
C(1)	9279(2)	4466(1)	440(2)	17(1)	C(20)	5437(2)	6431(1)	7445(2)	17(1)
C(2)	10585(2)	4368(1)	2280(2)	21(1)	C(21)	5437(2)	6814(1)	8171(2)	21(1)
C(3)	9682(2)	4711(1)	2463(2)	20(1)	C(22)	6233(2)	7170(1)	7722(2)	23(1)
C(4)	10998(2)	3829(1)	416(2)	21(1)	C(23)	7041(2)	7150(1)	6564(2)	23(1)
C(5)	10292(2)	3424(1)	970(2)	32(1)	C(24)	7015(2)	6765(1)	5834(2)	23(1)
C(6)	7676(2)	5089(1)	1101(2)	21(1)	C(25)	6217(2)	6406(1)	6269(2)	19(1)
C(7)	9496(2)	3609(1)	5710(2)	17(1)	C(26)	7957(3)	7530(1)	6096(3)	35(1)

Table 87: Atomic coordinates (x 10^4) and equivalent isotropic displacement parameters ($Å^2$ x 10^3) for [N$_{1,1,1,1}$][OTfa]. U(eq) is defined as one third of the trace of the orthogonalized U$_{ij}$ tensor.

	x	y	z	U(eq)
O(1)	1536(1)	1165(1)	6538(1)	24(1)
C(1)	513(3)	2500	8594(2)	26(1)
C(2)	1307(2)	2500	7056(1)	15(1)
F(1)	-1810(40)	2270(30)	8672(15)	46(3)
F(2)	1740(50)	1130(30)	9324(19)	45(4)
F(3)	1010(50)	3550(30)	9320(20)	31(2)
F(1B)	-2310(50)	2500	8460(30)	35(3)
F(2B)	1210(50)	1110(30)	9360(20)	26(4)
N(1)	-5067(2)	2500	3132(1)	15(1)
C(3)	-3535(3)	2500	1873(2)	21(1)
C(4)	-3380(3)	2500	4449(1)	18(1)
C(5)	-6664(2)	1053(1)	3110(1)	20(1)

Table 88: Atomic coordinates (x 10^4) and equivalent isotropic displacement parameters ($Å^2$ x 10^3) for [S$_{1,\varphi,\varphi}$][PF$_6$]. U(eq) is defined as one third of the trace of the orthogonalized U$_{ij}$ tensor.

	x	y	z	U(eq)		x	y	z	U(eq)
S(1)	3603(1)	6373(1)	4129(1)	20(1)	C(13)	5631(2)	8223(2)	5337(2)	26(1)
P	1441(1)	8727(1)	7173(1)	26(1)	F(1)	-211(6)	7560(12)	5767(4)	55(2)
C(1)	1309(2)	8448(2)	1318(2)	29(1)	F(2)	2648(9)	8319(8)	6440(9)	36(1)
C(2)	535(2)	7834(2)	2240(2)	30(1)	F(3)	1937(12)	10322(9)	6715(11)	60(2)
C(3)	1238(2)	7169(2)	3076(2)	25(1)	F(4)	250(11)	9126(10)	7921(10)	56(2)
C(4)	2728(2)	7170(2)	2980(1)	19(1)	F(5)	960(10)	7123(7)	7659(8)	36(1)
C(5)	3529(2)	7788(2)	2065(2)	24(1)	F(6)	3106(8)	9867(11)	8596(6)	38(2)
C(6)	2793(2)	8425(2)	1228(2)	28(1)	F(1A)	-243(7)	7866(17)	5789(5)	80(3)
C(7)	4274(2)	5210(2)	3069(1)	19(1)	F(2A)	2493(11)	8308(11)	6265(11)	60(3)
C(8)	5981(2)	5550(2)	3469(2)	26(1)	F(3A)	2114(13)	10503(10)	6939(14)	48(2)
C(9)	6364(2)	4501(2)	2641(2)	32(1)	F(4A)	409(8)	9192(8)	8081(8)	34(2)
C(10)	5078(2)	3177(2)	1438(2)	33(1)	F(5A)	775(15)	6984(10)	7410(16)	64(2)
C(11)	3381(2)	2865(2)	1046(2)	34(1)	F(6A)	3148(8)	9621(16)	8550(7)	36(2)
C(12)	2959(2)	3872(2)	1865(2)	27(1)					

6. Abbreviations

∠	angle	ppm	parts per million
a, b, c	unit cell dimensions	r_m	molecular radius
br	broad	RT	room temperature
d	distance/ day(s)	s	strong
DMF	$HC(O)N(CH_3)_2$	S	entropy
et al.	and others	$\hat{S}$	solvent-accessible surface
G	Gibbs free energy	sym.	symmetry/ symmetrical
h	hour(s)	t	time
H	enthalpy	T	temperature
HOMO	highest occupied molecular orbital	U	energy
IL	ionic liquid	VBT	volume-based thermodynamics
IR	infrared	vdW	van-der-Waals
J	coupling constant	V_m	molecular volume
L	ligand	vs	very strong
LUMO	lowest unoccupied molecular orbital	VSEPR	valence shell electron pair repulsion
m	medium/ multiplett	vw	very weak
M	molecular mass	w	weak
M	(metallic) central atom	WCA	weakly coordinating anion
min	minute(s)	α, β, γ	unit cell angles
MTBE	$CH_3OC(CH_3)_3$	γ, δ, ν, τ	stretch, in-plane deformation, out-of-plane deformation, torsion
NMR	nuclear magnetic resonance	ε_r	static dielectric constant
R	residue	$\tilde{\nu}$	wave number

7. General Conclusion and Outlook

The computational methods and correlations presented in this study may become helpful tools to make predictions to narrow down the number of syntheses needed, saving money, materials, time, and effort. We will now conclude this study with a recapitulation of these methods and their results.

7.1. Weakly Coordinating Anions

The predictive screening of possible new WCAs revealed some very promising compounds: $[Cu(hfip)_4]^-$, $[Au(dfc^6)_4]^-$, and $[Al(pfn^5)_4]^-$ (see Figure 101) show interesting properties and may be targets for future syntheses; once prepared, these WCAs could be used to stabilize very reactive and/or strongly coordinating cations.

We also showed that there may be pathways for decomposition with fluoride abstraction and the resulting destruction of a ligand which should also be taken into consideration when designing a WCA for a specific task. For example, the $[Au(pfn^5)_4]^-$ ion (see Figure 15) would not easily be synthesizable, since its possible precursor, $Au(pfn^5)_3$, is such a strong Lewis acid that it cleaves one ligand to saturate the central atom (assuming that no dimer is formed).

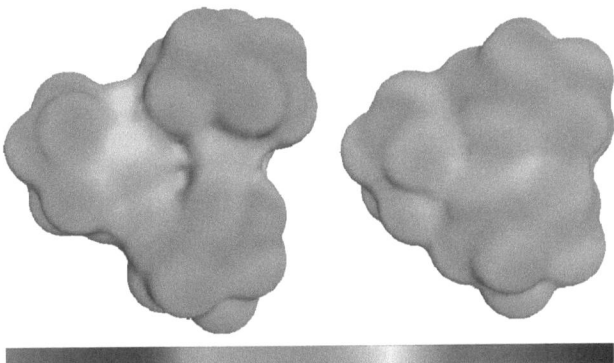

Figure 101: Left side: σ-surface of the most potent of all theoretically investigated WCAs, $[Al(pfn^5)_4]^-$; right side: σ-surface of $[Al(pftb)_4]^-$ for comparison. The color scale at the bottom shows the increasing degree of basicity. Pictures rendered with COSMOtherm.

7.2. Ionic Liquids

A new method for calculating the molecular volume was found, based on simple gas phase optimizations, discarding the need for the determination of crystal structures as the basis for all volume-based thermodynamics. The excellent accordance of calculated and measured X-ray volumes is reiterated in Figure 60 below.

Also, semi-empirical rules describing heat capacity, vitrification and melting point, critical micelle concentration, temperature-dependent density and liquid entropy as well as the basic thermodynamic phase change enthalpies of ionic liquids were discovered.

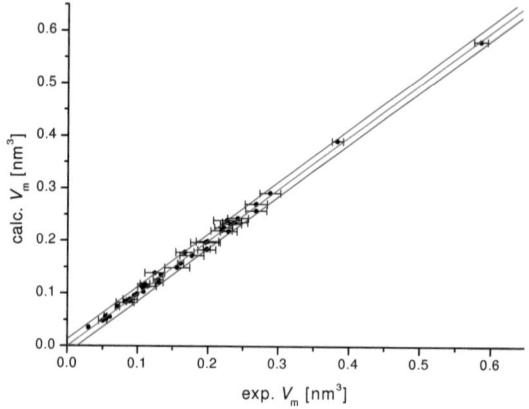

Figure 60: The connection of experimental and calculated (BP86/TZVP+COSMO) molecular volumes, shown for all anions and cations from Table 39 and Table 40 alongside the error bars and 95% prediction bands.

As already suggested in the prologue, the weak bonding of ionic liquids means that a purely statistical treatment – without taking directed interactions into account – suffices to accurately describe many properties. This means especially that in most cases, no ion pairs would need to be calculated, although the strongest non-covalent interactions are to be found in them. Even more astonishingly, the chemical nature of the single ions seems to be of little significance as well, since often, the most important contributions to our new prediction methods are traditional size-related quantities: the molecular volume or radius and the solvent-accessible surface. Correlations where the dependence from the molecular volume was hitherto not known are depicted in Figure 102. For melting point, density, vaporization and lattice enthalpy, new and improved formulas were discovered; for the standard liquid entropy, a new parametrization explicitly for ionic liquids was made.

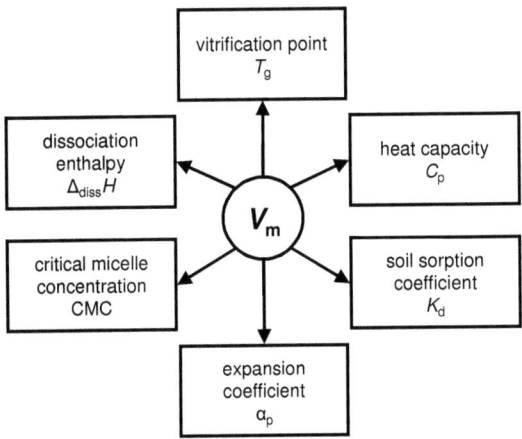

Figure 102: New dependencies of physical quantities on V_m delineated in this study.

For comparison, several quantities which were investigated in this study are listed along with their contributing factors in Table 89. Specific interactions, especially the ones calculated with COSMO(-RS), are often just used for broadening the applicable range as well as refining the results. One noteworthy exception is $\Delta_{latt}H^0$, surely "the" classical VBT quantity, which is almost completely expressed through $\Delta_{solv}H^{0,\infty}$ (findings reiterated in Figure 67 below).

Table 89: Selected quantities investigated in this study and their descriptors sorted by type.

quantity	descriptor			eqn.
	volume-based	thermodynamic	COSMO(-RS)	
α_p	$\ln(r_m)$	E_{ZP}		(16)
C_p^0	V_m			(17)
S_l^0	r_m^3			(25)
$\Delta_{vap}H^0$	$V_m^{2/3}$	H_g^0		(31)
$\Delta_{latt}H^0, \Delta_{diss}H^0$	$r_m^3, \hat{S}$	H_g^0	$\Delta_{solv}H^{0,\infty}$	(42), (44)
$\Delta_{fus}H$	r_m^3		H_{vdw}^0, H_{ring}	(53)
CMC	$r_m^3, \hat{S}$		$H_l^{diff}/\hat{S}, H_{ring}$	(67)

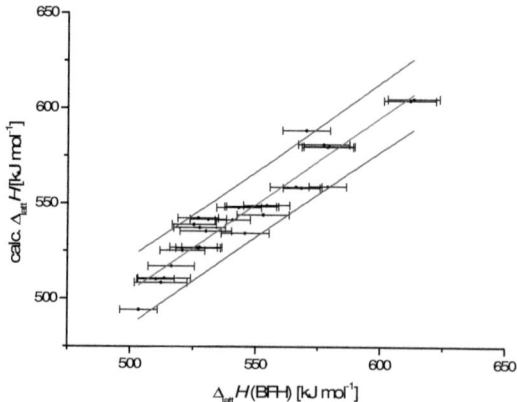

Figure 67: The correlation of lattice enthalpies taken from eqn. (27) (the BFH cycle) and (42), shown along with error bars and the 95% prediction bands.

Symmetry and conformational freedom play a huge role in the prediction of melting points, comprising the phase change entropy. The gas phase enthalpy plays a role in new equations for vaporization, dissociation, and lattice enthalpy, also bearing information about conformational freedom not accounted for in traditional VBT-style formulas. Much related is the influence of the zero-point energy on the expansion coefficient, which becomes crucial for compounds with long alkyl chains.

Originally, VBT formulas were parametrized for simple, inorganic salts with more or less point-shaped charges. In this context, the theoretical study of ionic liquids may prove not only a valuable tool in itself, but may also provide insight in thermodynamical principles that become apparent in complex molecules. In conclusion, we suggest the expression "aVBT" (augmented volume-based thermodynamics) to describe this kind of theoretical approach.

8. References

[1] V. Fock, *Z. Phys. A* **1930**, *61*, 126.

[2] D. R. Hartree, *Proc. Cambr. Phil. Soc.* **1928**, *24*, 111.

[3] P. L. L. Rosado, QTOctave 0.7.4, available at https://forja.rediris.es/projects/csl-qtoctave.

[4] A. D. Becke, *Phys. Rev. A* **1988**, *38*, 3098.

[5] A. Schäfer, H. Horn, R. Ahlrichs, *J. Chem. Phys.* **1992**, *97*, 2571.

[6] S. Vosko, L. Wilk, M. Nusair, *Can. J. Phys.* **1980**, *58*, 1200.

[7] J. P. Perdew, *Phys. Rev. B* **1986**, *33*, 8822.

[8] TURBOMOLE V5.10 **2008**, a development of University of Karlsruhe, 1989-2007, TURBOMOLE GmbH, since 2007 available from http://www.turbomole.com.

[9] F. Weigend, M. Häser, *Theor. Chem. Acc.* **1997**, *97*, 331.

[10] P. Deglmann, F. Furche, *J. Chem. Phys.* **2002**, *117*, 9535.

[11] P. Deglmann, F. Furche, R. Ahlrichs, *Chem. Phys. Lett.* **2002**, *362*, 511.

[12] A. Schäfer, C. Huber, R. Ahlrichs, *J. Chem. Phys.* **1994**, *100*, 5829.

[13] A. Klamt, G. Schüürmann, *J. Chem. Soc. Perkin Trans. 2*, **1993**, *5*, 799.

[14] F. Eckert, A. Klamt, COSMOtherm Version C2.1, Release 01.06, COSMOlogic GmbH & Co. KG, Leverkusen, Germany, **2006**.

[15] F. Eckert, A. Klamt, *AIChE J.* **2002**, *48*, 369.

[16] S. Renner, C. H. Schwab, G. Schneider, J. Gasteiger, *J. Chem. Inf. Model.* **2006**, *46*, 2324.

[17] The conformer generator ROTATE is available from Molecular Networks GmbH, Erlangen, Germany (http://www.molecular-networks.com).

[18] D. L. Theobald, D. S. Wuttke, *Bioinformatics* **2006**, *22*, 2171.

[19] I. Krossing, H. Brands, R. Feuerhake, S. Koenig, *J. Fluorine Chem.* **2001**, *112*, 83.

[20] I. Krossing, *Chem. Eur. J.* **2001**, *7*, 490.

[21] E. Bernhardt, G. Henkel, H. Willner, G. Pawelke, H. Burger, *Chem. Eur. J.* **2001**, *7*, 4696.

[22] E. Bernhardt, M. Finze, H. Willner, C. W. Lehmann, F. Aubke, *Angew. Chem. Int. Ed.* **2003**, *42*, 2077.

[23] L. Miinea, S. Suh, S. G. Bott, J.-R. Liu, W.-K. Chu, D. M. Hoffman, *J. Mater. Chem.* **1999**, *9*, 929.

[24] P.-L. Saaidi, E. Jeanneau, J. Hasserodt, *Acta Cryst. E* **2007**. *63*, o3241.

[25] C. Campbell, S. G. Bott, R. Larsen, W. G. Van Der Sluys, *Inorg. Chem.* **1994**, *33*, 4950.

[26] J. A. Samuels, E. B. Lobkovsky, W. E. Streib, K. Folting, J. C. Huffman, J. W. Zwanziger, K. G. Caulton, *J. Am. Chem. Soc.* **1993**, *115*, 5093.

[27] a) T. Endo, T. Kato, K.-I. Tozaki, K. Nishikawa, *J. Phys. Chem. B* **2010**, *114*, 407; b) K. Nishikawa, S. Wang, T. Endo, K.-I. Tozaki, *Bull. Chem. Soc. Jpn.* **2009**, *82*, 806; c) K. Nishikawa, S. Wang, K.-I. Tozaki, *Chem. Phys. Lett.* **2008**, *458*, 88.

[28] a) S. A. Katsyuba, T. P. Griaznova, A. Vidiš, P. J. Dyson, *J. Phys. Chem. B* **2009**, *113*, 5046; b) S. A. Katsyuba, E. E. Zvereva, A. Vidiš, P. J. Dyson, *J. Phys. Chem. A* **2007**, *111*, 352.

[29] F. Castiglione, M. Moreno, G. Raos, A. Famulari, A. Mele, G. B. Appetecchi, S. Passerini, *J. Phys. Chem. B* **2009**, *113*, 10750.

[30] V. N. Emel'yanenko, S. P. Verevkin, A. Heintz, C. Schick, *J. Phys. Chem. A* **2008**, *112*, 8095.

[31] A. V. Blokhin, Y. U. Paulechka, A. A. Strechan, G. J. Kabo, *J. Phys. Chem. B* **2008**, *112*, 4357.

[32] S. Zahn. B. Kirchner, *J. Phys. Chem. A* **2008**, *112*, 8430.

[33] T. Köddermann, C. Wertz, A. Heintz, R. Ludwig, *ChemPhysChem* **2006**, *7*, 1944.

[34] a) P. A. Hunt, B. Kirchner, T. Welton, *Chem. Eur. J.* **2006**, *12*, 6762; b) P. A. Hunt, I. R. Gould, *J. Phys. Chem. A* **2006**, *110*, 2269.

[35] Y. Wang, H. Li, S. Han, *J. Chem. Phys.* **2005**, *123*, 174501.

[36] S. Tsuzuki, H. Tokuda, K. Hayamizu, M. Watanabe, *J. Phys. Chem. B* **2005**, *109*, 16474.

[37] M. Bühl, A. Chaumont, R. Schurhammer, G. Wipff, *J. Phys. Chem. B* **2005**, *109*, 18591.

[38] S. Kunsági-Máté, B. Lemli, G. Nagy, L. Kollár, *J. Phys. Chem. B* **2004**, *108*, 9246.

[39] E. A. Turner, C. C. Pye, R. D. Singer, *J. Phys. Chem. A* **2003**, *107*, 2277.

[40] Z. Meng, A. Dölle, W. R. Carper, *J. Mol. Struct.* **2002**, *585*, 119.

[41] R. S. Mulliken, *J. Chem. Phys.* **1955**, *23*, 1833.

[42] R. Bauernschmitt, R. Ahlrichs, *Chem. Phys. Lett.* **1996**, *256*, 454.

[43] a) M. M. Francl, W. J. Petro, W. J. Hehre, J. S. Binkley, M. S. Gordon, D. J. DeFrees, J. A. Pople, *J. Chem. Phys.* **1982**, *77*, 3654; b) P. C. Hariharan, J. A. Pople, *Theor. Chim. Acta* **1973**, *28*, 213.

[44] L. A. Curtiss, P. C. Redfern, K. Raghavachari, V. Rassolov, J. A. Pople, *J. Chem. Phys.* **1999**, *110*, 4703.

[45] Gaussian 03, Revision C.02, M. J. Frisch, G. W. Trucks, H. B. Schlegel, G. E. Scuseria, M. A. Robb, J. R. Cheeseman, J. A. Montgomery, Jr., T. Vreven, K. N. Kudin, J. C. Burant, J. M. Millam, S. S. Iyengar, J. Tomasi, V. Barone, B. Mennucci, M. Cossi, G. Scalmani, N. Rega, G. A. Petersson, H. Nakatsuji, M. Hada, M. Ehara, K. Toyota, R. Fukuda, J. Hasegawa, M. Ishida, T. Nakajima, Y. Honda, O. Kitao, H. Nakai, M. Klene, X. Li, J. E. Knox, H. P. Hratchian, J. B. Cross, V. Bakken, C. Adamo, J. Jaramillo, R. Gomperts, R. E. Stratmann, O. Yazyev, A. J. Austin, R. Cammi, C. Pomelli, J. W. Ochterski, P. Y. Ayala, K. Morokuma, G. A. Voth, P. Salvador, J. J. Dannenberg, V. G. Zakrzewski, S. Dapprich, A. D. Daniels, M. C. Strain, O. Farkas, D. K. Malick, A. D. Rabuck, K. Raghavachari, J. B. Foresman, J. V. Ortiz, Q. Cui, A. G. Baboul, S. Clifford, J. Cioslowski, B. B. Stefanov, G. Liu, A. Liashenko, P. Piskorz, I. Komaromi, R. L. Martin, D. J. Fox, T. Keith, M. A. Al-Laham, C. Y. Peng, A. Nanayakkara, M. Challacombe, P. M. W. Gill, B. Johnson, W. Chen, M. W. Wong, C. Gonzalez, and J. A. Pople, Gaussian, Inc., Wallingford CT, **2004**.

[46] A. I. Yanovsky, Z. A. Starikova, N. Ya. Turova, D. E. Chebukov, E. P. Turevskaya, *Zh. Neorg. Khim.* **2002**, *47*, 1800.

[47] S. Worl, D. Hellwinkel, H. Pritzkow, M. Hofmann, R. Kramer, *Dalton Trans.* **2004**, 2750.

[48] Y. S. Vygodskii, O. A. Mel'nik, E. I. Lozinskaya, A. S. Shaplov, I. A. Malyshkina, N. D. Gavrilova, K. A. Lyssenko, M. Yu. Antipin, D. G. Golovanov, A. A. Korlyukov, N. Ignat'ev, U. Welz-Biermann, *Polym. Adv. Technol.* **2007**, *18*, 50.

[49] M. Jansen, H. Seyeda, *Z. Kristall. – New Cryst. Struc.* **1997**, *212*, 229.

[50] T. C. W. Mak, *J. Inclusion Phenom. Mol. Recog. Chem.* **1986**, *4*, 273.

[51] K. Miyatake, K. Oyaizu, Y. Nishimura, E. Tsuchida, *Macromol.* **2001**, *34*, 1172.

[52] I. Krossing, I. Raabe, *Angew. Chem. Int. Ed. Engl.* **2004**, *43*, 2066.

[53] I. Krossing, A. Reisinger, *Coord. Chem. Rev.* **2006**, *250*, 2721.

[54] U. Preiss, I. Krossing, *Z. Anorg. Allg. Chem.* **2007**, *633*, 1639.

[55] W. Dukat, D. Naumann, *Rev. Chim. Miner.* **1986**, *23*, 589.

[56] D. Naumann, T. Roy, K.-F. Tebbe, W. Crump, *Angew. Chem.* **1993**, *105*, 1555.

[57] J. A. Schlueter, J. M. Williams, U. Geiser, J. D. Dudek, S. A. Sirchio, M. E. Kelly, J. S. Gregar, W. H. Kwok, J. A. Fendrich, J. E. Schirber, W. R. Bayless, D. Naumann, T. Roy, *Chem. Commun.* **1995**, 1311.

[58] J. A. Schlueter, U. Geiser, A. M. Kini, H. H. Wang, J. M. Williams, D. Naumann, T. Roy, B. Hoge, R. Eujen, *Coord. Chem. Rev.* **1999**, 781, 190.

[59] I. Krossing, I. Raabe, *Chem. Eur. J.* **2004**, *10*, 5017.

[60] T. Roy, *PhD Thesis*: Cologne **1993**.

[61] H. Jenkins, D. Brooke, I. Krossing, J. Passmore, I. Raabe, *J. Fluorine Chem.* **2004**, *125*, 1585.

[62] W. Dukat, *PhD Thesis*: Dortmund **1986**.

[63] M. Finze, E. Bernhardt, M. Zähres, H. Willner, *Inorg. Chem.* **2004**, *43*, 490.

[64] However the reaction is endothermic in the gas phase, and only the high lattice energy of the precipitating K[AsF$_6$] makes the reaction favorable.

[65] D. Naumann, W. Tyrra, F. Trinius, W. Wessel, T. Roy, *J. Fluorine Chem.* **2000**, *101*, 131.

[66] It should be noted that Cu$^+$ only serves as a model for a soft electrophile and that CuF is not known; however, we believe it is a suitable model to investigate these reactions.

[67] a) C. Møller, M. S. Plesset, *Phys. Rev.* **1934**, *46*, 618; b) F. Weigend, M. Häser, H. Patzelt, R. Ahlrichs, *Chem. Phys. Lett.* **1998**, *294*, 143.

[68] K. Eichkorn, F. Weigend, O. Treutler, R. Ahlrichs, *Theor. Chem. Acc.* **1997**, *97*, 119.

[69] R. Ahlrichs, M. Bär, M. Häser, H. Horn, C. Kölmel, *Chem. Phys. Lett.* **1989**, *162*, 165.

[70] D. Andrae, U. Häussermann, M. Dolg, H. Stoll, H. Preuss, *Theor. Chim. Acta* **1990**, *77*, 123.

[71] a) E. R. Davidson, *J. Chem. Phys.* **1967**, *46*, 3320; b) K. R. Roby, *Mol. Phys.* **1974**, *27*, 81; c) C. Ehrhardt, R. Ahlrichs, *Theor. Chim. Acta* **1985**, *68*, 231.

[72] L. O. Müller, *PhD Thesis*: Freiburg **2008**.

[73] L. O. Müller, D. Himmel, J. Stauffer, G. Steinfeld, J. Slattery, G. Santiso-Quiñones, V. Brecht, I. Krossing, *Angew. Chem. Int. Ed.* **2008**, *47*, 7659.

[74] L. O. Müller, R. Scopelliti, I. Krossing, *Chimia* **2006**, *60*, 220.

[75] T. Okazoe, K. Watanabe, M. Itoh, D. Shirakawa, H. Murofushi, H. Okamoto, S. Tatematsua, *Adv. Synth. Catal.* **2001**, *343*, 215.

[76] P. Walden, *Bull. Acad. Imper. Sci.* **1914**, 405.

[77] P. Bonhote, A.-P. Dias, N. Papageorgiou, K. Kalyanasundaram, M. Grätzel, *Inorg. Chem.* **1996**, *35*, 1168.

[78] J. S. Wilkes, M. J. J. Zaworotko, *Chem. Commun.* **1992**, 965.

[79] J. S. Wilkes, J. A. Levisky, R. A. Wilson, C. L. Hussey, *Inorg. Chem.* **1982**, *21*, 1263.

[80] P. Wasserscheid, T. Welton, *Ionic Liquids in Synthesis*, Wiley-VCH: Weinheim, **2008**.

[81] F. Endres, A. Abbott, D. R. MacFarlane, *Electrodeposition in Ionic Liquids*; Wiley-VCH: Weinheim, **2008**.

[82] The *Green Chem.* issue 2 in **2002**, the *Acc. Chem. Res.* issue 11 in **2007**, the *Austr. J. Chem.* issue 1 in **2007**, the *J. Phys. Chem. B* issue 18 in **2007**, the *Monatsh. Chem.* issue 11 in **2007**, the *Int. J. Mol. Sci.* Special Issue "Ionic Liquids" (**2007** – **2008**), and the *J. Organomet. Chem.* issue 15 in **2010**, for example, are dedicated to many aspects of IL work.

[83] H. Weingärtner, *Angew. Chem. Int. Ed.* **2008**, *47*, 654.

[84] G. Drake, T. Hawkins, A. Brand, L. Hall, M. Mckay, A. Vij, I. Ismail, *Prop., Explos., Pyrotech.* **2003**, *28*, 174.

[85] J. D. Holbrey, K. R. Seddon, *Clean Prod. Proc.* **1999**, *1*, 223.

[86] F. Endres, S. Z. El Abedin, *Phys. Chem. Chem. Phys.* **2006**, *8*, 2101.

[87] J. M. Slattery, C. Daguenet, P. J. Dyson, T. J. S. Schubert, I. Krossing, *Angew. Chem. Int. Ed.* **2007**, *46*, 5384.

[88] I. Krossing, J. M. Slattery, C. Daguenet, P. J. Dyson, A. Oleinikova, H. Weingärtner, *J. Am. Chem. Soc.* **2006**, *128*, 13427.

[89] C. Ye, J. M. Shreeve, *J. Phys. Chem. A* **2007**, *111*, 1456.

[90] I. Krossing, J. M. Slattery, *Z. Phys. Chem.* **2006**, *220*, 1343.

[91] M. Deetlefs, K. R. Seddon, M. Shara, *Phys. Chem. Chem. Phys.* **2006**, *8*, 642.

[92] A. P. Abbott, *ChemPhysChem* **2005**, *6*, 2502.

[93] A. R. Katritzky, R. Jain, A. Lomaka, R. Petrukhin, M. Karelson, A. E. Visser, R. D. Rogers, *J. Chem. Inf. Comput. Sci.* **2002**, *42*, 225.

[94] S. Trohalaki, R. Pachter, *QSAR Comb. Sci.* **2005**, *24*, 485.

[95] S. Trohalaki, R. Pachter, S. R. Drake, T. Hawkins, *Energy & Fuels* **2005**, *19*, 279.

[96] A. P. Abbott, *ChemPhysChem* **2004**, *5*, 1242.

[97] D. M. Eike, J. F. Brennecke, E. J. Maginn, *Green Chem.* **2003**, *5*, 323.

[98] S. Alavi, D. L. Thompson, *J. Chem. Phys.* **2005**, *122*, 154704.

[99] A. R. Katritzky, A. Lomaka, R. Petrukhin, R. Jain, M. Karelson, A. E. Visser, R. D. Rogers, *J. Chem. Inf. Comput. Sci.* **2002**, *42*, 71.

[100] Charles Augustin de Coulomb, 1736 – 1806, French physicist; Johannes Diderik van der Waals, 1837 – 1923, Dutch physicist.

[101] C. Chiappe, *Monatsh. Chem.* **2007**, *138*, 1035.

[102] a) D. Xiao, L. G. Hines, S. Li, R. A. Bartsch, E. L. Quitevis, O. Russina, A. Triolo, *J. Phys. Chem. B* **2009**, *113*, 6426; b) L. Gontrani, O. Russina, F. Lo Celso, R. Caminiti, G. Annat, A. Triolo, *J. Phys. Chem. B* **2009**, *113*, 9235; c) A. Triolo, O. Russina, B. Fazio, G. B. Appetecchi, M. Carewska, S. Passerini, *J. Chem. Phys.* **2009**, *130*, 164521; d) A. Triolo, O. Russina, B. Fazio, R. Triolo, E. Di Cola, *Chem. Phys. Lett.* **2008**, *457*, 362; e) A. Triolo, O. Russina, H.-J. Bleif, E. Di Cola, *J. Phys. Chem. B* **2007**, *111*, 4641.

[103] J. N. A. Canongia Lopes, A. A. H. Pádua, *J. Phys. Chem. B* **2006**, *110*, 3330.

[104] a) Y. Wang, G. A. Voth, *J. Phys. Chem. B* **2006**, *110*, 18601; b) Y. Wang, G. A. Voth, *J. Am. Chem. Soc.* **2005**, *127*, 12192.

[105] C. Schröder, T. Rutas, O. Steinhauser, *J. Chem. Phys.* **2006**, *125*, 244506.

[106] J. de Andrade, E. S. Boes, H. Stassen, *J. Phys. Chem. B* **2008**, *112*, 8966.

[107] J. Habasaki, K. L. Ngai, *J. Chem. Phys.* **2008**, *129*, 194501.

[108] B. L. Bhargava, R. Devane, M. L. Klein, S. Balasubramanian, *Soft Matter* **2007**, *3*, 1395.

[109] O. Borodin, G. D. Smith, *J. Phys. Chem. B* **2006**, *110*, 11481.

[110] a) C. Hardacre, J. D. Holbrey, M. Nieuwenhuyzen, T. G. A. Youngs, *Acc. Chem. Res.* **2008**, *40*, 1146; b) C. Hardacre, S. E. J. McMath, M. Nieuwenhuyzen, D. T. Bowron, A. K. Soper, *J. Phys. Condens. Matter* **2003**, *15*, S159; c) C. Hardacre, J. D. Holbrey, S. E. J. McMath, D. T. Bowron, A. K. Soper, *J. Chem. Phys.* **2003**, *118*, 273.

[111] Q. L. Kuang, J. Zhang, Z. G. Wang, *J. Phys. Chem. B* **2007**, *111*, 9858.

[112] S. Shigeto, H. Hamaguchi, *Chem. Phys. Lett.* **2006**, 329.

[113] D. Xiao, J. R. Rajian, S. Li, R. A. Bartsch, E. L. Quitevis, *J. Phys. Chem. B* **2006**, *110*, 16174.

[114] A.-L. Rollet, P. Porion, M. Vaultier, I. Billard, M. Deschamps, C. Bessarda, L. Jouvensal, *J. Phys. Chem. B* **2007**, *111*, 11888.

[115] W. Jiang, Y. Wang, G. A. Voth, *J. Phys. Chem. B* **2007**, *111*, 4812.

[116] C. G. Hanke, R. M. Lynden-Bell, *J. Phys. Chem. B* **2003**, *107*, 10873.

[117] N. Sieffert, G. Wipff, *J. Phys. Chem. A* **2006**, *110*, 1106.

[118] a) A. A. H. Pádua, M. F. Costa Gomes, J. N. A. Canongia Lopes, *Acc. Chem. Res.* **2007**, *40*, 1087; b) J. N. Canongia Lopes, M. F. Costa Gomes, A. A. H. Padua, *J. Phys. Chem. B* **2006**, *110*, 16816.

[119] G. Chevrot, R. Schurhammer, G. Wipff, *Phys. Chem. Chem. Phys.* **2006**, *8*, 4166.

[120] J. A. Widegren, A. Laesecke, J. W. Magee, *Chem. Commun.* **2005**, *12*, 1610.

[121] K. R. Seddon, A. Stark, M.-J. Torres, *Pure Appl. Chem.* **2000**, *72*, 2275.

[122] E. F. Pettersen, T. D. Goddard, C. C. Huang, G. S. Couch, D. M. Greenblatt, E. C. Meng, T. E. Ferrin, *J. Comput. Chem.* **2004**, *25*, 1605.

[123] Persistence of Vision Pty. Ltd. (2004): Persistence of Vision Raytracer (Version 3.6), retrieved from http://www.povray.org/download

[124] A. Pádua, Thermodynamics 2005 Conference, Sesimbra, Portugal, **2005**; Congress on Ionic Liquids, Salzburg, Austria, **2005**.

[125] U. Preiss, J. M. Slattery, I. Krossing, *Ind. Eng. Chem. Res.* **2009**, *48*, 2290.

[126] L. Glasser, H. D. B. Jenkins, *Chem. Soc. Rev.* **2005**, *34*, 866.

[127] L. Glasser, H. D. B. Jenkins, *Thermochim. Acta* **2004**, *414*, 125.

[128] H. D. B. Jenkins, L. Glasser, *Inorg. Chem.* **2003**, *42*, 8702.

[129] H. D. B. Jenkins, H. K. Roobottom, J. Passmore, L. Glasser, *Inorg. Chem.* **1999**, *38*, 3609.

[130] T. E. Mallouk, G. L. Rosenthal, G. Muller, R. Brusasco, N. Bartlett, *Inorg. Chem.* **1984**, *23*, 3167.

[131] L. P. N. Rebelo, J. N. Canongia Lopes, J. M. S. S. Esperança, H. J. R. Guedes, J. Łachwa, V. Najdanovic-Visak, Z. P. Visak, *Acc. Chem. Res.* **2007**, *40*, 1114.

[132] J. M. S. S. Esperança, H. J. R. Guedes, M. Blesic, L. P. N. Rebelo, *J. Chem. Eng. Data* **2006**, *51*, 237.

[133] L. P. N. Rebelo, V. Najdanovic-Visak, R. Gomes de Azevedo, J. M. S. S. Esperança, M. Nunes da Ponte, H. J. R. Guedes, H. C. de Sousa, J. Szydlowski, J. N. Canongia Lopes, T. C. Cordeiro, in: R. D. Rogers, K. R. Seddon, *Ionic Liquids IIIA: Fundamentals, Progress, Challenges, and Opportunities-Properties and Structure* American Chemical Society: Washington DC, **2005**.

[134] D. W. M. Hofmann, *Acta Cryst.* **2002**, *B57*, 489.

[135] H. L. Ammon, *Struct. Chem.* **2001**, *12*, 205.

[136] C. M. Tarver, *J. Chem. Eng. Data* **1979**, *24*, 136.

[137] J. C. McGowan, *Rec. Trav. Chim. Pays-Bas* **1956**, *75*, 193.

[138] Y. Marcus, *Biophys. Chem.* **2006**, *124*, 200.

[139] L. Pauling, *The nature of the chemical bond*, Cornell University Press: New York, 1960, 3rd edition.

[140] A. Gavezzotti, *J. Am. Chem. Soc.* **1983**, *105*, 5220.

[141] K. E. Gutowski, J. D. Holbrey, R. D. Rogers, D. A. Dixon, *J. Phys. Chem. B* **2005**, *109*, 23196.

[142] It should be noted that the standard mode of Gaussian 03 used to calculate the volumes gives rather scattering values of the volume (10 % error, see ref. [141] and an experienced user is needed to change the grid used for integration.

[143] D. Tran, J. P. Hunt, S. Wherland, *Inorg. Chem.* **1992**, *31*, 2460.

[144] J. R. Witt, D. Britton, *Acta Cryst. B*, **1971**, *27*, 1835.

[145] P. Andersen, B. Klewe, E. Thom, *Acta Chem. Scand.* **1967**, *21*, 1530.

[146] P. Andersen, B. Klewe, *Nature* **1963**, *200*, 464.

[147] H. D. B. Jenkins, J. F. Liebman, *Inorg. Chem.* **2005**, *44*, 6359.

[148] I. Raabe, *PhD Thesis*: Freiburg, **2007**.

[149] J. J. P. Stewart, *J. Comput. Chem.* **1989**, *10*, 209.

[150] J. J. P. Stewart, *J. Mol. Model.* **2007**, *13*, 1173.

151 A. Klamt, F. Eckert, *Fluid Phase Equilib.* **2000**, *172*, 43.

152 A. Bondi, *J. Phys. Chem.* **1964**, *68*, 441.

153 J. B. Foresman, T. A. Keith, K. B. Wiberg, J. Snoonian, M. J. Frisch, *J. Phys. Chem.* **1996**, *100*, 16098.

154 As in every correlation, r^2 cannot be affected by multiplication or addition, if uniformly applied to all calculated x- or y-values therefore, the presence or absence of a linear fit does not change it.

155 J. J. P. Stewart, Stewart Computational Chemistry, Colorado Springs, CO, USA, 2007. URL: http://openmopac.net.

156 B. C. Taverner, University of the Witwatersrand, South Africa, **1995** version 1.11.

157 R. L. Gardas, J. A. P. Coutinho, *Fluid Phase Equilib.* **2008**, *263*, 26.

158 J. Jacquemin, R. Ge, P. Nancarrow, D. W. Ronney, M. F. Costa Gomes, A. A. H. Pádua, C. Hardacre, *J. Chem. Eng. Data* **2008**, *53*, 716.

159 K. Binnemans, *Chem. Rev.* **2005**, *105*, 4148.

160 J. Troncoso, C. A. Cerdeirina, Y. A. Sanmamed, L. Romani, L. P. N. Rebelo, *J. Chem. Eng. Data* **2006**, *51*, 1856.

161 A. B. Pereiro, F. Santamarta, E. Tojo, A. Rodríguez, J. Tojo, *J. Chem. Eng. Data* **2006**, *51*, 952.

162 K. R. Harris, L. A. Woolf, M. Kanakubo, *J. Chem. Eng. Data* **2005**, *50*, 1777.

163 G. J. Kabo, A. V. Blokhin, Y. U. Paulechka, A. G. Kabo, M. P. Shymanovich, *J. Chem. Eng. Data* **2004**, *49*, 453.

164 Z. Gu, J. F. Brennecke, *J. Chem. Eng. Data* **2002**, *47*, 339.

165 A. Wandschneider, J. K. Lehmann, A. Heintz, *J. Chem. Eng. Data* **2008**, *53*, 596.

166 J. Jacquemin, P. Husson, V. Mayer, I. Cibulka, *J. Chem. Eng. Data* **2007**, *52*, 2204.

167 K. R. Harris, M. Kanakubo, L. A. Woolf, *J. Chem. Eng. Data* **2007**, *52*, 1080.

168 E. Gómez, B. González, N. Calvar, E. Tojo, Á. Domínguez, *J. Chem. Eng. Data* **2006**, *51*, 2096.

169 J. Jacquemin, P. Husson, A. A. H. Padua, V. Majer, *Green Chem.* **2006**, *8*, 172.

170 E. Gómez, B. González, Á. Domínguez, E. Tojo, J. Tojo, *J. Chem. Eng. Data* **2006**, *51*, 696.

171 A. A. Strechan, Y. U. Paulechka, A. V. Blokhin, G. J. Kabo, *J. Chem. Thermodyn.* **2008**, *40*, 632.

172 A. T. M. Golam Mostafa, J. M. Eakman, M. M. Montoya, S. L. Yarbro, *Ind. Eng. Chem. Res.* **1996**, *35*, 343.

173 Y. U. Paulechka, G. J. Kabo, A. V. Blokhin, A. S. Shaplov, E. I. Lozinskaya, Ya. S. Vygodskii, *J. Chem. Thermodyn.* **2007**, *39*, 158.

174 L. P. N. Rebelo, V. Najdanovic-Visak, Z. P. Visak, M. Nunes da Ponte, J. Szydlowski, C. A. Cerdeirina, J. Troncoso, L. Romani, J. M. S. S. Esperanca, H. J. R. Guedes, H. C. de Sousa, *Green Chem.* **2004**, *6*, 369.

175 URL (accessed 2007-Dec-18): http://ilthermo.boulder.nist.gov/ILThermo/pureprp.uix.do

176 C. P. Fredlake, J. M. Crosthwaite, D. G. Hert, S. N. V. K. Aki, J. F. Brennecke, *J. Chem. Eng. Data* **2004**, *49*, 954.

177 J. M. Crosthwaite, M. J. Muldoon, J. K. Dixon, J. L. Anderson, J. F. Brennecke, *J. Chem. Thermodyn.* **2005**, *37*, 559.

[178] A. V. Blokhin, Y. U. Paulechka, G. J. Kabo, *J. Chem. Eng. Data* **2006**, *51*, 1377.

[179] D. Waliszewski, I. Stepniak, H. Piekarski, A. Lewandowski, *Thermochim. Acta* **2005**, *433*, 149.

[180] K.-S. Kim, B.-K. Shin, H. Lee, F. Ziegler, *Fluid Phase Equilib.* **2004**, *218*, 215.

[181] Z.-H. Zhang, Z.-C. Tan, L.-X. Sun, J.-Z. Yang, X.-C. Lv, Q. Shi, *Thermochim. Acta* **2006**, *447*, 141.

[182] U. Preiss, V. N. Emel'yanenko, S. P. Verevkin, D. Himmel, Y. U. Paulechka, I. Krossing, *ChemPhysChem* **2010**, accepted.

[183] M. Diedenhofen, A. Klamt, K. Marsh, A. Schäfer, *Phys. Chem. Chem. Phys.* **2007**, *33*, 4653.

[184] F. Eckert, COSMOtherm Users Manual, Version C2.1, Release 01.06, COSMOlogic GmbH & Co. KG, Leverkusen, Germany, **2006**.

[185] Y. U. Paulechka, A. V. Blokhin, G. J. Kabo, A. A. Strechan, *J. Chem. Thermodynam.* **2007**, *39*, 866.

[186] A. A. Strechan, A. G. Kabo, Y. U. Paulechka, A. V. Blokhin, G. J. Kabo, A. S. Shaplov, E. I. Lozinskaya, *Thermochim. Acta* **2008**, *474*, 25.

[187] A. A. Strechan, Y. U. Paulechka, A. G. Kabo, A. V. Blokhin, G. J. Kabo, *J. Chem. Eng. Data* **2007**, *52*, 1791.

[188] R. D. Chirico, V. Diky, J. W. Magee, M. Frenkel, K. M. Marsh, *Pure Appl. Chem.* **2009**, *81*, 791.

[189] a) M. Born, *Verhandl. Deut. Physik. Ges.* **1919**, *21*, 679; b) K. Fajans, *Verhandl. Deut. Physik. Ges.* **1919**, *21*, 714; c) F. Haber, *Verhandl. Deut. Physik. Ges.* **1919**, *21*, 750.

[190] P. Wasserscheid, *Nature* **2006**, *439*, 797.

[191] M. J. Earle, J. M. S. S. Esperança, M. A. Gilea, J. N. Canongia Lopes, L. P. N. Rebelo, J. W. Magee, K. R. Seddon, J. A. Widegren, *Nature* **2006**, *439*, 831.

[192] R. Bini, O. Bortolini, C. Chiappe, D. Pieraccini, T. Siciliano, *J. Phys. Chem. B* **2007**, *111*, 598.

[193] J. M. L. Martin, G. De Oliveira, *J. Chem. Phys.* **1999**, *111*, 1843.

[194] A. D. Boese, A. Chandra, J. M. L. Martin, D. Marx, *J. Chem. Phys.* **2003**, *119*, 5965.

[195] S. Zahn, F. Uhlig, J. Thar, C. Spickermann, B. Kirchner, *Angew. Chem. Int. Ed.* **2008**, *47*, 3639.

[196] S. Grimme, *J. Comput. Chem.* **2006**, *27*, 1787.

[197] C. Lee, W. Yang, R. G. Parr, *Phys. Rev. B* **1988**, *37*, 785.

[198] M. Gerenkamp, S. Grimme, *Chem. Phys. Lett.* **2004**, *392*, 229.

[199] Dr. S. Verevkin and Dr. V. N. Emel'yanenko, University of Rostock, private communication.

[200] a) G. E. Scuseria, C. L. Janssen, H. F. Schaefer III, *J. Chem. Phys.* **1988**, *89*, 7382; b) G. D. Purvis III, R. J. Bartlett, *J. Chem. Phys.* **1982**, *76*, 1910; c) J. A. Pople, R. Krishnan, H. B. Schlegel, J. S. Binkley, *Int. J. Quantum Chem.* **1978**, *14*, 545; d) P. C. Hariharan, *Advances in Chemical Physics*, Wiley Interscience: New York, **1969**.

[201] a) D. E. Woon, T. H. Dunning Jr., *J. Chem. Phys.* **1993**, *98*, 1358; b) R. A. Kendall, T. H. Dunning Jr., R. J. Harrison, *J. Chem. Phys.* **1992**, *96*, 6796; c) T. H. Dunning Jr., *J. Chem. Phys.* **1989**, *90*, 1007.

[202] J. P. Armstrong, C. Hurst, R. G. Jones, P. Licence, K. R. J. Lovelock, C. J. Satterley, I. J. Villar-Garcia, *Phys. Chem. Chem. Phys.* **2007**, *9*, 982.

203 J. M. S. S. Esperança, J. N. Canongia Lopes, M. Tariq, L. M. N. B. F. Santos, J. W. Magee, L. P. N. Rebelo, *J. Chem. Eng. Data* **2010**, *55*, 3.

204 T. Köddermann, D. Paschek, R. Ludwig, *ChemPhysChem* **2008**, *9*, 549.

205 M. Martín-Betancourt, J. M. Romero-Enrique, L. F. Rull, *J. Phys. Chem. B* **2009**, *113*, 9046.

206 P. Ballone, C. Pinilla, J. Kohanoff, M. G. Del Pópolo, *J. Phys. Chem. B* **2007**, *111*, 4938.

207 D. H. Zaitsau, G. J. Kabo, A. A. Strechan, Y. U. Paulechka, A. Tschersich, S. P. Verevkin, A. Heintz, *J. Phys. Chem. A* **2006**, *110*, 7303.

208 V. N. Emel'yanenko, S. P. Verevkin, A. Heintz, *J. Am. Chem. Soc.* **2007**, *129*, 3930.

209 S. P. Verevkin, *Angew. Chem. Int. Ed.* **2008**, *47*, 5071.

210 Y. U. Paulechka, Dz. H. Zaitsau, G. J. Kabo, A. A. Strechan, *Thermochim. Acta* **2005**, *439*, 158.

211 A. Deyko, K. J. R. Lovelock, J.-A. Corfield, A. W. Taylor, P. N. Gooden, I. J. Villar-Garcia, P. Licence, R. G. Jones, V. G. Krasovskiy, E. A. Chernikova, L. M. Kustov, *Phys. Chem. Chem. Phys.* **2009**, *11*, 8544.

212 K. Rakus, S. P. Verevkin, J. Schätzer, H.-D. Beckhaus, C. Rüchardt, *Chem. Ber.* **1994**, *127*, 1095.

213 H. Luo, G. A. Baker, S. Dai, *J. Phys. Chem. B* **2008**, *112*, 10077.

214 Here, direct calculation with COSMO-RS would give too low a value also (153 kJ mol^{-1}). The paramagnetism of the compound likely influences the spin density, yielding a higher solvation energy and therefore a higher vaporization enthalpy.

215 A. Jain, S. H. Yalkowsky, *J. Pharm. Sci.* **2006**, *95*, 2562.

216 L. Zhao, S. H. Yalkowsky, *Ind. Eng. Chem. Res.* **1999**, *38*, 3581.

217 a) N. V. Sidgwick, *The Covalent Link in Chemistry*, Cornell University Press: Ithaca, NY, **1933**; b) T. L. Cottrell, *The Strengths of Chemical Bonds*, Butterworths: London, **1954**.

218 M. E. Van Valkenburg, R. L. Vaughn, M. Williams, J. S. Wilkes, *Thermochim. Acta* **2005**, *425*, 181.

219 D. R. MacFarlane, P. Meakin, N. Amini, M. Forsyth, *J. Phys. Condens. Matter* **2001**, *13*, 8257.

220 P. Wasserscheid, R. van Hal, A. Bosmann, *Green Chem.* **2002**, *4*, 400.

221 H. Tokuda, K. Hayamizu, K. Ishii, M. A. B. H. Susan, M. Watanabe, *J. Phys. Chem. B* **2005**, *109*, 6103.

222 U. Domanska, P. Morawski, *Green Chem.* **2007**, *9*, 361.

223 U. Domanska, A. Marciniak, *J. Chem. Eng. Data* **2003**, *48*, 451.

224 H. Sifaoui, A. Ait-Kaci, A. Modarressi, M. Rogalski, *Thermochim. Acta* **2007**, *456*, 114.

225 U. Domanska, Z. Zolek-Tryznowska, M. Krolikowski, *J. Chem. Eng. Data* **2007**, *52*, 1872.

226 J. S. Chickos, S. Hosseini, D. G. Hesse, J. F. Liebman, *Struct. Chem.* **1993**, *4*, 271.

227 T. Nishida, Y. Tashiro, M. Yamamoto, *J. Fluorine Chem.* **2003**, *120*, 135.

228 D. S. H. Wong, J. P. Chen, J. M. Chang, C. H. Chou, *Fluid Phase Equilib.* **2002**, *194 – 197*, 1089.

229 H. Tokuda, S. Tsuzuki, M. A. B. H. Susan, K. Hayamizu, M. Watanabe, *J. Phys. Chem. B* **2006**, *110*, 19593.

230 D. M. Camaioni, C. A. Schwerdtfeger, *J. Phys. Chem. A* **2005**, *109*, 10795.

231 H. D. B. Jenkins, K. F. Pratt, *Proc. R. Soc.* **1977**, *356*, 115.

[232] A. F. Kapustinskii, Z. Phys. Chem. B **1933**, *22*, 257.
[233] K. E. Gutowski, R. D. Rogers, D. A. Dixon, *J. Phys. Chem. B* **2007**, *111*, 4788.
[234] U. Preiss, S. Bulut, I. Krossing, *J. Phys. Chem. B* **2010**, *114*, 11133.
[235] R. Bini, C. Chiappe, C. Duce, A. Micheli, R. Solaro, A. Starita, M. R. Tiné, *Green Chem.* **2008**, *10*, 306.
[236] A. Varnke, N. Kireeva, *J. Chem. Inf. Model.* **2007**, *47*, 1111.
[237] I. López-Martin, E. Burello, P. N. Davey, K. R. Seddon, G. Rothenberg, *ChemPhysChem* **2007**, *8*, 690.
[238] N. Sun, X. He, K. Dong, X. Zhang, X. Lu, H. He, S. Zhang, *Fluid Phase Equilib.* **2006**, *246*, 137.
[239] G. Carrera, J. Aires-de-Sousa, *Green Chem.* **2005**, *7*, 20.
[240] A. R. Katritzky, V. S. Lobanov, M. Karelson, University of Florida **1994**.
[241] G. V. S. M. Carrera, L. C. Branco, J. Aires-de-Sousa, C. A. M. Afonso, *Tetrahedron* **2008**, *64*, 2216.
[242] W. Zhao, F. Leroy, B. Heggen, S. Zahn, B. Kirchner, S. Balasubramanian, F. Müller-Plathe, *J. Am. Chem. Soc.* **2009**, *131*, 15825.
[243] R. Ludwig, *Phys. Chem. Chem. Phys.* **2008**, *10*, 4333.
[244] W. Zhao, H. Eslami, W. L. Cavalcanti, F. Müller-Plathe, *Z. Phys. Chem.* **2007**, *221*, 1647.
[245] T. Köddermann, D. Paschek, R. Ludwig, *ChemPhysChem* **2007**, *8*, 2464.
[246] C. Cadena, Q. Zhao, R. Q. Snurr, E. J. Maginn, *J. Phys. Chem. B* **2006**, *110*, 2821.
[247] J. N. Canongia Lopes, A. A. H. Pádua, *J. Phys. Chem. B* **2004**, *108*, 16893.
[248] Z. Liu, S. Huang, W. Wang, *J. Phys. Chem. B* **2004**, *108*, 12978.
[249] C. Margulis, *J. Mol. Phys.* **2004**, *102*, 829.
[250] E. J. Maginn, T. I. Morrow, *J. Phys. Chem. B* **2002**, *106*, 12807.
[251] C. J. Margulis, H. A. Stern, B. J. Berne, *J. Phys. Chem. B* **2002**, *106*, 12017.
[252] C. G. Hanke, S. L. Price, R. M. Lynden-Bell, *Mol. Phys.* **2001**, *99*, 801.
[253] J. Fuller, R. T. Carlin, H. C. De Long, D. Haworth, *Chem. Commun.* **1994**, 299.
[254] J. Solca, A. J. Dyson, G. Steinebrunner, B. Kirchner, B. Huber, *J. Chem. Phys.* **1998**, *108*, 4107.
[255] J. Solca, A. J. Dyson, G. Steinebrunner, B. Kirchner, H. Huber, *Chem. Phys.* **1997**, *224*, 253.
[256] J. F. Lutsko, D. Wolf, S. R. Phillpot, S. Yip, *Phys. Rev. B* **1989**, *40*, 2841.
[257] S. Jayaraman, E. J. Maginn, *J. Chem. Phys.* **2007**, *127*, 214504.
[258] R.-M. Dannenfelser, S. H. Yalkowsky, *Ind. Eng. Chem. Res.* **1996**, *35*, 1483.
[259] M. A. Neumann, F. J. J. Leusen, J. Kendrick, *Angew. Chem. Int. Ed.* **2008**, *47*, 2427.
[260] L. Glasser, *Thermochim. Acta* **2004**, *421*, 87.
[261] a) A. König, M. Stepanski, A. Kuszlik, P. Keil, C. Weller, *Chem. Eng. Res. Design* **2008**, *86*, 775; b) Z. Fei, W. H. Ang, D. Zhao, R. Scopelliti, E. E. Zvereva, S. A. Katsyuba, P. J. Dyson, *J. Phys. Chem. B* **2007**, *111*, 10095.
[262] K. R. J. Lovelock, A. Deyko, J.-A. Corfield, P. N. Gooden, P. Licence, R. G. Jones, *ChemPhysChem* **2009**, *10*, 337.
[263] B. R. Clare, P. M. Bayley, A. S. Best, M. Forsyth, D. R. MacFarlane, *Chem. Commun.* **2008**, 2689.

[264] L. G. Bonnet, B. M. Kariuki, *Eur. J. Inorg. Chem.* **2006**, 437.
[265] I. Raabe, K. Wagner, K. Guttsche, M. Wang, M. Grätzel, G. Santiso-Quinoñes, I. Krossing, *Chem. Eur. J.* **2009**, *15*, 1966.
[266] S. Bulut, *PhD Thesis:* Freiburg, **2010**.
[267] J. W. Gibbs, *Am. J. Sci.* **1878**, *16*, 441.
[268] J. Bohm, *Kristallzüchtung*, Verlag Harri Deutsch: Frankfurt/Main **1988**.
[269] U. Preiss, C. Jungnickel, J. Thöming, I. Krossing, J. Luczaka, M. Diedenhofen, A. Klamt, *Chem. Eur. J.* **2009**, *15*, 8880.
[270] M. Herstedt, W. A. Henderson, M. Smirnov, L. Ducasse, L. Servant, D. Talaga, J. C. Lassègues, *J. Mol. Struct.* **2006**, *783*, 145.
[271] J. D. Holbrey, W. M. Reichert, R. D. Rogers, *Dalton Trans.* **2004**, 2267.
[272] T. Timofte, S. Pitula, A.-V. Mudring, *Inorg. Chem.* **2007**, *46*, 10938.
[273] J. Sun, M. Forsyth, D. R. MacFarlane, *J. Phys. Chem. B* **1998**, *102*, 8858.
[274] H. Matsumoto, H. Kageyama, Y. Miyazaki, *Chem. Lett.* **2001**, *30*, 182.
[275] H. L. Ngo, K. LeCompte, L. Hargens, A. B. McEwen, *Thermochim. Acta* **2000**, *357 – 358*, 97.
[276] H. Matsumoto, H. Kageyama, Y. Miyazaki, *Chem. Commun.* **2002**, 1726.
[277] Z.-B. Zhou, H. Matsumoto, K. Tatsumi, *Chem. Eur. J.* **2005**, *11*, 752.
[278] D. MacFarlane, *J. Phys. Cond. Matter* **2001**, *13*, 8257.
[279] J. Zhang, G. R. Martin, D. D. DesMarteau, *Chem. Commun.* **2003**, 2334.
[280] Taken from the Solvionic SA product catalogue, URL (accessed 2009-Jun-23): http://www.solvionic.com/index.php?context=82
[281] S. K. Quek, I. M. Lyapkalo, H. V. Huynh, *Tetrahedron* **2006**, *62*, 3137.
[282] J. O'M. Bockris, A. K. N. Reddy, *Modern Electrochemistry*, 2nd edition, Plenum Press: New York **1998**.
[283] C. J. Margulis, *Mol. Phys.* **2004**, *102*, 829.
[284] A. L. Greer, *Nature* **2000**, *404*, 145.
[285] W. Xu, E. I. Cooper, C. A. Angell, *J. Phys. Chem. B* **2003**, *107*, 6170.
[286] W. Xu, L.-M. Wang, R. A. Nieman, C. A. Angell, *J. Phys. Chem. B* **2003**, *107*, 11749.
[287] F. J. Baltá Calleja, D. S. Sanditov, V. P. Privalko, *J. Mater. Sci.* **2002**, *37*, 4507.
[288] T. Erdmenger, J. Vitz, F. Wiesbrock, U. S. Schubert, *J. Mater. Chem.* **2008**, *18*, 5267.
[289] A. Cavagna, *Phys. Rep.* **2009**, *476*, 51.
[290] H. Zhao, *J. Chem. Technol. Biotechnol.* **2006**, *81*, 877.
[291] A. Fernandez, J. S. Torrecilla, F. Rodriquez, *J. Chem. Eng. Data* **2007**, *52*, 1979.
[292] J. G. Huddleston, A. E. Visser, W. M. Reichert, H. D. Willauer, G. A. Broker, R. D. Rogers, *Green Chem.* **2001**, *3*, 156.
[293] J. Dudowicz, K. F. Freed, J. F. Douglas, *J. Phys. Chem. B* **2005**, *109*, 21285.
[294] U. Domanska, A. Rekawek, A. Marciniak, *J. Chem. Eng. Data* **2008**, *53*, 1126.

[295] I. Gutzow, S. Ilieva, F. Babalievski, V. Yamakov, *J. Chem. Phys.* **2000**, *112*, 10941.

[296] V. R. Thalladi, R. Boese, H.-C. Weiss, *J. Am. Chem. Soc.* **2000**, *122*, 1186.

[297] V. R. Thalladi, R. Boese, H.-C. Weiss, *Angew. Chem. Int. Ed.* **2000**, *39*, 918.

[298] D. G. Archer, *National Insitute of Standards and Technology NISTIR 6645*, **2006**.

[299] R. P. Wool, *J. Polymer Sci. B* **2008**, 2765.

[300] G. M. Bartenev, *Dokl. Akad. Nauk SSSR*, **1951**, *76*, 227.

[301] a) C. T. Moynihan, A. J. Easteal, M. A. DeBolt, J. Tucker, *J. Am. Ceram. Soc.* **1976**, *59*, 12; b) C. T. Moynihan, A. J. Easteal, M. A. DeBolt, J. Tucker, *J. Am. Ceram. Soc.* **1976**, *59*, 16.

[302] A. B. Bainova, D. S. Sanditov, *Glass Phys. Chem.* **2002**, *28*, 189.

[303] G. M. Bartenev, I. A. Luk'yanov, *Zh. Fiz. Khim.* **1955**, *29*, 1486.

[304] W. M. Reichert, J. D. Holbrey, R. P. Swatloski, K. E. Gutowski, A. E. Visser, M. Nieuwenhuyzen, K. R. Seddon, R. D. Rogers, *Crystal Growth & Design*, **2007**, *7*, 1106.

[305] T. Ruether, J. Huang, A. F. Hollenkamp, *Chem. Commun.* **2007**, *48*, 5226.

[306] Y. Abu-Lebdeh, A. Abouimrane, P.-J. Alarco, M. Armand, *J. Power Sources* **2006**, *154*, 255.

[307] S. A. Forsyth, K. J. Fraser, P. C. Howlett, D. R. MacFarlane, M. Forsyth, *Green Chem.* **2006**, *8*, 256.

[308] P.-J. Alarco, Y. Abu-Lebdeh, M. Armand, *Solid State Ionics*, **2004**, *175*, 717.

[309] Technical Summaries on Ionic Liquids in Chemical Processing, in Vision 2020, Oak Ridge National Laboratory, USA, **2003**.

[310] C. Jungnickel, J. Łuczak, J. Ranke, J. F. Fernández, A. Müller, J. Thöming, *Colloids Surf. A* **2008**, *316*, 278.

[311] Z. Miskolczy, K. Sebok-Nagy, L. Biczok, S. Gokturk, *Chem. Phys. Lett.* **2004**, *400*, 296.

[312] M. Blesic, M. H. Marques, N. V. Plechkova, K. R. Seddon, L. P. N. Rebelo, A. Lopes, *Green Chem.* **2007**, *9*, 481.

[313] J. Bowers, C. P. Butts, P. J. Martin, M. C. Vergara-Gutierrez, *Langmuir* **2004**, *20*, 2191.

[314] R. Vanyur, L. Biczok, Z. Miskolczy, *Colloids Surf. A* **2007**, *299*, 256.

[315] J. Sirix-Plenet, L. Gaillon, P. Letellier, *Talanta* **2004**, *63*, 979.

[316] Q. Q. Baltazar, J. Chandawalla, K. Sawyer, J. L. Anderson, *Colloids Surf. A* **2007**, *302*, 150.

[317] O. V. Vieira, D. O. Hartmann, C. M. P. Cardoso, D. Oberdoerfer, M. Baptista, M. A. S. Santos, L. Almeida, J. Ramalho-Santos, W. L. C. Vaz, *PLoS ONE* **2008**, *3*, e2913.

[318] P. D. T. Huibers, V. S. Lobanov, A. R. Katritzky, D. O. Shah, M. Karelson, *Langmuir* **1996**, *12*, 1462.

[319] R. Nagarajan, E. Ruckenstein, *Langmuir* **1991**, *7*, 2934.

[320] U. P. R. M. Preiss, J. M. Slattery, I. Krossing, *Ind. Eng. Chem. Res.* **2009**, *48*, 2290.

[321] G. D'Errico, O. Ortona, L. Paduano, V. Vitagliano, *J. Colloid Interface Sci.* **2001**, *239*, 264.

[322] W. Mosquera, *J. Colloid Interface Sci.* **1998**, *206*, 66.

[323] T. M. Perger, M. Bester-Rogac, *J. Colloid Interface Sci.* **2007**, *313*, 288.

[324] I. Chakraborty, S. P. Moulik, *J. Phys. Chem. B* **2007**, *111*, 3658.

325 S. K. Mehta, K. K. Bhasin, R. Chauhan, S. Dham, *Colloids Surf. A* **2005**, *255*, 153.

326 P. Hansson, B. Joensson, C. Stroem, O. Soederman, *J. Phys. Chem. B* **2000**, *104*, 3496.

327 Y. Hayami, H. Ichikawa, A. Someya, M. Aratono, K. Motomura, *Colloid Polym. Sci.* **1998**, *276*, 595.

328 S. Durand-Vidal, M. Jardat, V. Dahirel, O. Bernard, K. Perrigaud, P. Turq, *J. Phys. Chem. B* **2006**, *110*, 15542.

329 R. Ueoka, Y. Murakami, *J. Chem. Soc. Perkin Trans. 2* **1983**, 219.

330 I. Goodchild, L. Collier, S. L. Millar, I. Prokeš, J. C. D. Lord, C. P. B. Butts, J. Bowers, J. R. P. Webster, R. K. Heenan, *J. Colloid Interface Sci.* **2007**, *307*, 445.

331 J. Wang, H. Wang, S. Zhang, H. Zhang, Y. Zhao, *J. Phys. Chem. B* **2007**, *111*, 6181.

332 O. A. El Seoud, P. A. R. Pires, T. Abdel-Moghny, E. L. Bastos, *J. Colloid Interface Sci.* **2007**, *313*, 296.

333 B. Dong, N. Li, L. Zheng, L. Yu, T. Inoue, *Langmuir* **2007**, *23*, 4178.

334 T. Inoue, H. Ebina, B. Dong, L. Zheng, *J. Colloid Interface Sci.* **2007**, *314*, 236.

335 S. Thomaier, K. Werner, *J. Mol. Liq.* **2007**, *130*, 104.

336 F. Quina, *J. Phys. Chem.* **1995**, *99*, 17028.

337 T. Sasaki, M. Hattori, J. Sasaki, K. Nukina, *Bull. Chem. Soc. Jpn.* **1975**, *48*, 1397.

338 A. Modaressi, H. Sifaoui, M. Mielcarz, U. Domanska, M. Rogalski, *Colloids Surf. A* **2007**, *302*, 181.

339 Z. Hu, C. J. Margulis, *Acc. Chem. Res.* **2007**, *40*, 1097.

340 P. Stepnowski, W. Mrozik, J. Nichthauser, *Environ. Sci. Technol.* **2007**, *41*, 511.

341 A. A. Kornyshev, *J. Phys. Chem. B* **2007**, *111*, 5545.

342 J. B. Rollins, B. D. Fitchett, J. C. Conboy, *J. Phys. Chem. B* **2007**, *111*, 4990.

343 C. Jungnickel, M. Markiewicz, U. Preiss, W. Mrozik, P. Stepnowski, *J. Optoelectron. Adv. Mater. Symposia* **2009**, *1*, 82.

344 R. E. A. Dear, W. B. Fox, R. J. Fredericks, E. E. Gilbert, *Inorg. Chem.* **1970**, *9*, 2590.

345 J. J. Rockwell, G. M. Kloster, W. J. DuBay, P. A. Grieco, D. F. Shriver, S. H. Strauss, *Inorg. Chim. Acta* **1997**, *263*, 195.

346 J. L. Adcock, H. Zhang, *J. Org. Chem.* **1996**, *61*, 5073.

347 Dr. T. Sonoda, Kyushu University, Japan, private communication.

Die VDM Verlagsservicegesellschaft sucht für wissenschaftliche Verlage abgeschlossene und herausragende

Dissertationen, Habilitationen, Diplomarbeiten, Master Theses, Magisterarbeiten usw.

für die kostenlose Publikation als Fachbuch.

Sie verfügen über eine Arbeit, die hohen inhaltlichen und formalen Ansprüchen genügt, und haben Interesse an einer honorarvergüteten Publikation?

Dann senden Sie bitte erste Informationen über sich und Ihre Arbeit per Email an *info@vdm-vsg.de*.

Sie erhalten kurzfristig unser Feedback!

VDM Verlagsservicegesellschaft mbH
Dudweiler Landstr. 99
D - 66123 Saarbrücken

Telefon +49 681 3720 174
Fax   +49 681 3720 1749

www.vdm-vsg.de

Die VDM Verlagsservicegesellschaft mbH vertritt

Printed by Books on Demand GmbH, Norderstedt / Germany